THE CYDONIA CODEX

THE CYDONIA CODEX
Reflections from Mars

George Haas

William Saunders

Forewords by
Dr. Mark J. Carlotto and **Richard C. Hoagland**

Frog, Ltd.
Berkeley, California

Published by Frog, Ltd.
Frog, Ltd. books are distributed
by North Atlantic Books
P.O. Box 12327
Berkeley, California 94712

Cover design by Suzanne Albertson
Book design by Brad Greene
Printed in the United States of America
Distributed to the book trade by
Publishers Group West

North Atlantic Books' publications are available through most bookstores. For further information, call 800-337-2665 or visit our website at www.northatlanticbooks.com.

Substantial discounts on bulk quantities are available to corporations, professional associations, and other organizations. For details and discount information, contact our special sales department.

Library of Congress Cataloging-in-Publication Data

Haas, George J., 1957–
    The Cydonia codex : reflections from Mars / by George J. Haas and William R. Saunders ; foreword by Mark J. Carlotto ; foreword by Richard C. Hoagland.
        p. cm.
    Includes bibliographical references and index.
    Summary: "The result of ten years of study and analysis of NASA photographs of the Face on Mars and its surrounding complex, The Cydonia Codex provides evidence for a terrestrial connection between Cydonia and Mesoamerica"—Provided by publisher.
    ISBN 1-58394-121-5 (pbk.)
    1. Mars (Planet)—Surface—Photographs from space. 2. Cydonia (Mars) 3. Extraterrestrial anthropology. 4. Life on other planets. I. Saunders, William R., 1952– II. Title.
    QB641.H24 2005
    523.43—dc22

2005007079
CIP

1 2 3 4 5 6 7 8 9 DATA 10 09 08 07 06 05

The sacred symbols of the cosmic elements were hid away hard by the secrets of Osiris. Hermes, ere he returned to Heaven, invoked a spell on them and spake these words: "... O holy books, who have been made by my immortal hands, by incorruption's magic spells ... free from decay throughout eternity remain and incorrupt from time! Become unseeable, unfindable, for every one whose foot shall tread the plains of this land, until old Heaven doth bring forth meet instruments for you, whom the Creator shall call souls. Thus spake he; and, laying spells on them by means of his own works, he shut them safe away in their zones. And long enough the time has been since they were hid away." "Men will seek out ... the inner nature of the holy spaces which no foot may tread, and will chase after them into the height, desiring to observe the nature of the motion of the Heavens."

— Egyptian treatise, *The Virgin of the World*

The original book, written long ago existed, but its sight is hidden from the searcher and the thinker.

—The Maya *Popol Vuh*

Table of Contents

Acknowledgments

Advancements in understanding the extent of our existence on the planet Earth cannot be achieved without recognizing the great achievements by those who came before us. We are forever indebted to those who have made archival recordings of their time in history and those who now strive to interpret and understand the remnants of those precious records. So it is with great appreciation that we extend our sincere gratitude to all who have kept mankind's existence on Earth and beyond a sacred and living document of humanity.

The authors would also like to extend our sincere thanks to our publisher, Richard Grossinger, and the staff at Frog Ltd. (North Atlantic Books). A personal thanks is extended to Linda Heine and Paige Miller, who read and commented on earlier drafts of this work.

A special thanks goes to our colleagues Lee Bogart, George Dutton, and our new associate Barry Baumgardner for all of their perceptive comments and suggestions over the last five years of research. Many thanks go to Mark Carlotto, Keith Laney, and Richard Hoagland for granting permission to use their enhancements of *Viking* and *MGS* images. We also thank Jim Miller for his fine work in reproducing the transparent overlays of the Lid of Pacal by Maurice M. Cotterell. We are most grateful to Sylvia Perrine at FAMSI for allowing us to use the drawings of Linda Schele, and to Soichi Sunami at the Isamu Noguchi Foundation for the use of her photograph of Noguchi's sculpture. We particularly wish to thank JPL and NASA for providing the public with such an amazing archive of Martian images. We are also extremely indebted to Mary Dant at the Goddard Space Flight Center and gratefully thank her for her insights into the workings of NASA.

Finally, we would like to thank our loving families, who have wholeheartedly supported us while researching and writing this book.

Foreword

Dr. Mark J. Carlotto

It has long been recognized that, while the "Face on Mars" possesses a high degree of symmetry in terms of its overall structure, internally it is far from symmetrical. In the 1976 *Viking* orbiter images of the "Face," with the sun setting to the northwest, cast shadows and shading effects were initially thought to be responsible for its unbalanced appearance. Closer examination revealed that the distortions from symmetry on the right, partially shadowed, eastern side were not optical illusions but real. In large part, because the "Face" does not appear to be symmetrical today, planetary scientists have dismissed it as a natural formation. They believe that the random forces of nature have sculpted the "Face," like New Hampshire's Old Man of the Mountain and other terrestrial features.

It became clear from the April 2000 image taken by the *Mars Global Surveyor*, that the "Face" is not only asymmetrical but also highly eroded. This image provided some new clues as to why the right side of the structure does not match the left. It appears that a combination of erosion and deposition are responsible for what we see today—this was verified a year later in the almost overhead, fully illuminated April 2001 *MGS* image. There are indications that internal features have collapsed and become highly eroded, like the rest of the formation.

Computer-generated perspective views suggest the presence of sand dunes on the right side. It is possible that the dunes are the result of the prevailing winds scouring the western side of the "Face" and depositing the eroded material on the leeward side. It is my hypothesis that the "Face" is a highly symmetrical artificial structure that has been transformed into an eroded asymmetrical landform—not unlike the neighboring mesas and buttes in this part of Cydonia—as a result of its long-term exposure to the Martian environment.

Often, advances in science are the result of analyzing a new phenomenon from different perspectives, from seeing the same thing in different ways. *The Cydonia Codex* advances a different hypothesis from the one I have outlined above. Citing terrestrial analogs from Central America, George Haas and William Saunders argue that the "Face on Mars" is asymmetrical by design—depicting a different facial representation on each of its sides. In applying the scientific method to the study of an anomaly like the "Face," scientists often adhere to Ockham's principle—that the simplest explanation is the best one. For the mainstream science community, the simplest explanation is that the "Face" is only a funny-looking rock formation. But given the amount of evidence uncovered over the past quarter century in favor of artificiality, this explanation seems naïve. In analyzing the "Face" and other anomalies on Mars from our limited planetary perspective, we must be open to all plausible explanations. For, if we are not, we might miss an important clue in our search for the true origin of these enigmatic features on Mars.

—Mark Carlotto

August 2002

Dr. Mark J. Carlotto has published more than fifty papers on unusual objects on the planet Mars. His work has appeared in many science journals and in Carl Sagan's Cosmos television series. He is the author of *The Martian Enigmas* and *The Cydonia Controversy.*

Foreword

Richard C. Hoagland

For more than twenty years, NASA has insistently proclaimed that the celebrated "Face on Mars" is nothing but "a trick of light and shadow." The long-awaited 2001 release of a full-faced image of the "Face" (MOC image E03-00824) clearly refuted this parochial official view. It provided unequivocal evidence of the "Face" platform's strikingly artificial symmetry, as well as elegantly confirmed the highly provocative human/feline aspect of its visage—key predictions I had made at the United Nations over a decade before!

The Enterprise Mission has consistently maintained since we began our own investigations that if the "Face on Mars" was artificially created, it should be consistent with the whole of Cydonia—which displays many additional examples of this underlying "dual asymmetry," as well as an amazing geometric pattern. From a scientific point of view, it is this designed pattern that now overwhelmingly confirms the artificial nature of this complex, revealed in the highly repetitive, tetrahedral mathematics incorporated in Cydonia's various structures. It was this extraordinary geometric legacy that, early on, led us to expect that, if "the Builders" were indeed trying to say something profound with their mile-long, 1,500-foot-high "Face," it would be through this surrounding context—the pivotal, geometric "Message of Cydonia." It is just such a communicative geometry that ultimately formed the insight for my own Hyperdimensional Physics Model, extensively documented in *The Monuments of Mars*.

But, surprisingly, there now seem to be potential cultural answers to the enigma of Cydonia as well, beginning with the haunting duality of the "Face" itself—answers that may now hearken back to the most ancient of our cultures here on Earth—a link that I termed in *Monuments,* "the Terrestrial Connection."

For, if the "Face on Mars" is an authentic work of art, if the Cydonia Complex is reality, if this ancient Martian civilization once did in fact exist, one would ultimately have to get into the minds of its creators to truly understand their presence. You'd have to get to the root of what they intended to say by their mysterious creation of a gigantic, mile-long monument that looks like us . . . and something more.

In the pages that follow, the authors of *The Cydonia Codex* may have provided us with just such an important, long-awaited insight into some of the, till now, hidden reasoning behind "The Monuments of Mars." George Haas and William Saunders have found and published in this volume an extraordinary reservoir of two-faced artifacts, created within the major pre-Columbian cultures of this planet—beginning with Central America. Their findings neatly confirm via a totally independent terrestrial model the controversial, bifurcated aspects of the "Face on Mars" I first proposed in 1992. The Haas-Saunders research provides clear evidence that the Maya (among others) created the same symbolic duality in their sculpture, mythology, and writing system as we have now discovered in the remarkable hominid/feline symbolism of this ancient extraterrestrial "monument," lying where it has no business being . . . in a northern Martian region called Cydonia.

A crucial future task for all researchers will be to attempt to discover if the terrestrial mythologies associated with these remarkable dual images of Mesoamerica can be traced back to the point of human origins; and in turn, if they will ultimately lead us back across an almost unfathomable immensity of time . . . all the way to Mars. NASA, along with artists, historians, anthropologists, and geologists, must unite to connect these data points and test this intriguing new hypothesis—and soon. For, in the end, the answers may eventually lead us to a profound redefinition of literally who and what we are. The implications that are set forth in this research—and in the "Face" itself—are staggering, for just as Colonel John Wilder gazed into his own reflection in a fabled Martian canal in Ray Bradbury's classic *The Martian Chronicles,* so we may all soon realize that it is truly us who are the Martians after all.

—Richard C. Hoagland

August 2003

Richard Hoagland is founder and principal investigator of The Enterprise Mission, recipient of the Angstrom Medal, former science advisor to CBS News and Walter Cronkite, author of *The Monuments of Mars,* co-creator of the Pioneer Plaque, and originator of the Europa Proposal.

Introduction

George J. Haas and **William R. Saunders**

The controversial "Face on Mars" that has sparked the world's attention was discovered while the citizens of America were engaged in celebrating their nation's Bicentennial. As fireworks were "bursting in air" below the starlit skies of America, the *Viking 1* orbiter was snapping pictures of a mysterious red planet millions of miles away. Amazingly, one of these pictures, from an area called Cydonia, would reveal an anomalous mesa that exhibited a most remarkable portrait—of a humanoid face. It was with this one photograph that the real fireworks would ultimately begin.

Following its release, NASA firmly stated that the "Face" was nothing more than "a trick of light" and any thoughts to the contrary were quickly dismissed as utterly ridiculous. But it was too late. There was nothing NASA could do to turn back the approaching tide of public opinion. Around the world the controversy had already taken on a life of its own. This one single *Viking* photograph would soon become "the Face that launched a thousand conspiracies."

The notion that such a colossal edifice could have been built on Mars was truly beyond belief. However, many found it hard to conceive that such a facial structure could have occurred naturally. To the sculptor's eye, the "Face" appeared very lifelike. It had facial features that projected a distinctly mask-like quality and it wore what appeared to be some kind of helmet. Surely this haunting image could not be the result of an optical illusion—it had to be much more. Unfortunately, it would take more than twenty years before this so-called "trick of light" would begin to unveil its secrets to a volatile world.

The first opportunity for the public to explore the surface of Mars in any reasonable detail came in 1998 with the images from the Mars Orbital Camera (MOC) aboard the *Mars Global Surveyor* (*MGS*) spacecraft. It was with great anticipation and apprehension that concerned citizens of the world visited NASA's web site in an effort to downloaded their first high-resolution images of the surface of the red planet.

With three new images of the Cydonia area to examine, a hodgepodge of independent researchers began contributing to the conference-room discussion boards on Richard Hoagland's website, The Enterprise Mission. Most notable was the a new thread set up by Mark Archambault in anticipation of the release of the upcoming *MGS* images, which were expected to include new images of the "Face." The discussion was titled "Cydonia in April." It was in this very discussion room that the authors first met and began what would truly become a fantastic journey. Soon after reviewing a few of our core discoveries[1] and sharing our thoughts and insights about the anomalous structures found at Cydonia, we decided to join forces and "set the world on edge."

Within a month we set up a private website[2] to archive our research and solicit feedback from qualified researchers. As soon as our site was up and running, we contacted Richard Hoagland and Mark Carlotto to get their reactions to our discoveries. We received no response from either of them. We also notified two members of the Society for Planetary SETI Research (SPSR) and allowed them access to the site. Although the response here was quick, they showed very little interest in our observations.

In an effort to present our findings to a broader audience, we launched The Cydonia Institute's official website in October 2000.[3] We thought the site would allow high-profile researchers the opportunity to view our discoveries and make critical comments, thereby allowing us to build a reference file of endorsements. Unfortunately, our site never lasted long enough to be an effective tool. After little more than two months, our website mysteriously disappeared! It appeared somebody out there didn't like what we were finding! How little we knew of the power and perils of web exposure!

On the evening of May 25, 2001, a message from Richard C. Hoagland was left for us on The Cydonia Institute's answering machine. The all-too-familiar voice said: "Yeah, this is Dick Hoagland of The Enterprise Mission . . . . I need to speak to you as soon as practicable. My number here is . . . . Thanks a lot. Bye."

After ten years of research, someone was finally taking notice. We took a deep breath, and gave "Dick" a call.

Richard Hoagland explained that the reason he hadn't contacted us earlier was that he was awaiting confirmation of the feline side of the "Face," which would confirm his split-faced hypothesis. After reviewing the new 2001 "Face" image and realizing its cultural implications, he recalled our five-year-old conversation that alerted him to a common two-faced iconography used by Maya and Olmec cultures. Hoagland requested that we send him a few examples of those two-faced masks a.s.a.p., so we snapped into action and e-mailed him as much supportive material as we could. We also provided him with a link to our recently restored website.

That very evening, on the Art Bell "Coast to Coast" radio show, Richard Hoagland announced his recognition of a two-faced Mesoamerican connection to the "Face on Mars" and credited this observation to the findings of George Haas and William Saunders. It appeared our work was finally being noticed and would now be exposed to an audience that was large enough to make an impact.

The stunning discoveries that we have made on the surface of Mars and its relationship to our ancient past are revealed for the first time in the pages that follow. Our intent is to offer both the general public and the scientific community unprecedented evidence that NASA has photographed monumental edifices that parallel the sculpture and mythology of ancient cultures from Mesoamerica, Asia, Mesopotamia, and Egypt on the surface of Mars. The implication of this discovery is staggering: Earth's history and mankind's origins are far different than previously believed.

Notes

1. A 1998 manuscript by George J. Haas connecting the Face on Mars to the masks of Mesoamerica, titled "Two-Faced," was sent to Richard Hoagland, Mark Carlotto, Terry James (a.k.a. KKsamurai), Gary Leggiere (a.k.a. Mars Revealer), Chronos (a.k.a Robert Carl), Clayton Barr, and Baruch Cowley (among others) for review. After Terry James went public with the contents of the manuscript on his discussion board "The Mars Round-table," requests became so overwhelming that distribution of the manuscript was stopped. From that point on, all of our discoveries were kept confidential and reserved for publication only.

2. This private site was launched on May 3, 1998 and was shut down in February 1999.

3. The Cydonia Institute membership included George J. Haas, William R. Saunders, Lee Bogart, and George Dutton.

CHAPTER 1

The Face on Mars

The *Viking* Orbiter
(a Trick of Light and Shadow)

On July 25, 1976, the *Viking 1* orbiter, circling the planet Mars at an altitude of 1,000 miles, snapped the first picture of a mesa that had an incredible resemblance to a human face. This unusual structure was discovered among the initial cascade of pictures broadcast back to Earth from the spacecraft during the mapping sequence of the entire planet. This face-like mesa, approximately a mile and a half long and a mile wide, was initially spotted on *Viking* frame 35A72 by a member of National Aeronautics and Space Administration's (NASA's) own imaging team at the Jet Propulsion Laboratory (JPL) in Pasadena, California.

While searching for a possible landing site for the upcoming *Viking 2* lander in the Cydonia region, Dr. Tobias Owen noticed a gigantic, human-like face glaring up at him from the barren Martian surface. Michael Carr, who was then head of the *Viking* orbiter imaging team, immediately released the unusual image to the press.

Soon after its discovery, NASA spokesman Gerry Soffen announced at a press conference that an image of an odd landform resembling a face had been found. However, the press and news media were quickly informed that when a second picture was taken only a few hours later, the image of the "Face" had disappeared. Oddly, this second image never surfaced.

NASA's official position was that the "Face" was an apparition of shadows and rock, and the overall mesa had no resemblance to a face. Consequently, only a single high-contrast picture of the "Face" was circulated to the press as nothing more than a phantom novelty (Figure 1.1).

Conveniently, NASA decided not to proceed with the original plan of a *Viking 2* landing at Cydonia because the area was deemed unsafe. The *Viking 2* lander eventually set down in a rocky plain called Utopia. This last-

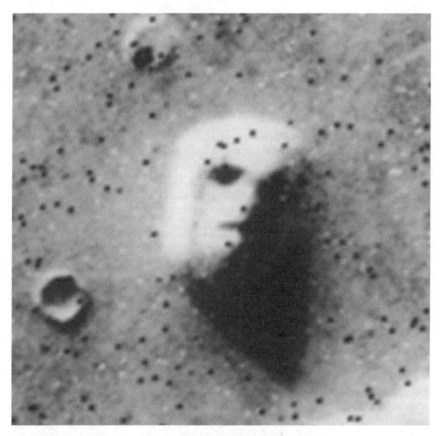

1.1 Original NASA *Viking* orbiter version of "Face on Mars" (frame 35A72).
Courtesy of NASA/JPL/Caltech

minute change in plans went virtually unquestioned by the media. With NASA's firm and consistent stance that the face-like landform was nothing more than a "trick of light and shadow," the public soon lost interest.

From that point on, the "Face on Mars" was banished to the sensational pages of supermarket tabloids and the illusion-filled minds of fringe-science enthusiasts. It would not be until three years later that an independent team of imaging specialists, Vincent DiPietro and Gregory Molenaar, discovered additional *Viking* images of the "Face" that had somehow been overlooked.

Vincent DiPietro was a computer scientist with NASA's Goddard Space Flight Center. His partner, Greg Molenaar, was a computer scientist with Lockheed Martin on contract with NASA at the Computer Sciences Corporation. The team discovered a second image of the "Face" that had been strangely "misfiled" and buried in the massive archives of images that were

taken by the *Viking* orbiter in 1976. How could this be? NASA had assured us that there was only one picture showing the image of the "Face" and that that unique picture was only a "trick of light and shadow." It appeared that NASA's original denial of the existence of any other pictures of the "Face" was now coming into question.

Not only did DiPietro and Molenaar find additional pictures of the "Face," they found at least one picture, frame 70A13 (Figure 1.2), with better resolution than the first. This second image was with a higher sun angle and, contrary to what NASA claimed, also resembles a human face.

By 1979, this team of pioneering imaging specialists had developed a method for enhancing

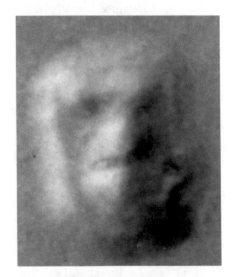

1.2 "Face on Mars" from NASA *Viking* frame 70A13. This image was taken 35 days after the image in frame 35A72.
Computer-enhanced photo courtesy of Dr. Mark J. Carlotto

the "Face" called Starburst Pixel Interleaving Technique (SPIT), capable of accessing more detail from the original *Viking* tapes.[1] After applying this technique to two different images of the "Face" and completing detailed measurements of its structure, they were convinced that the "Face" was real. The enhanced second image featured an obvious headdress and exposed more of the far side of the mesa, revealing an amazing symmetry. In 1980, DiPietro and Molenaar produced a monograph entitled *Unusual Martian Surface Features* and set out to enlighten the world with their findings: "Not only did the second frame confirm the first, but additional features emerged. The contour of the eye cavity became more distinct. The hairline continued to the opposite side."[2]

Following a small press conference and a few independent presentations of their state-of-the-art computer enhancements of the "Face" and surrounding structures, public and scientific opinion was unchanged. It seemed nobody (including NASA) was listening; it was too little, too late.[3]

DiPietro and Molenaar, continuing with their research, were joined by physicist Dr. John Brandenburg in 1985. While re-examining the available data they began a thorough search of the NASA files for additional *Viking* images of the "Face." After a relentless search, they discovered a data set of ten images that the *Viking* orbiter had shot over the area of the "Face."[4] Although most of the images were of poor quality, they substantiated NASA's lack of veracity. It was now quite obvious that more than one image of the "Face" was obtained.

In 1983 Richard Hoagland, at the time a science journalist, had taken up the investigation of Mars. He followed the fine work of DiPietro and Molenaar and eventually published a book entitled *The Monuments of Mars: A City on the Edge of Forever*. Hoagland put together a team of scientists from outside the NASA community called the Independent Mars Investigation to study the Cydonia area. Their work produced the idea of a message of hyperdimensional physics encoded into mathematical relationships between the "Face" and other Cydonia-area structures.

In 1984, Hoagland proclaimed the "Face on Mars" to be the embodiment of a Martian Sphinx,[5] where the face was half humanoid and half feline. The profound implication of Hoagland's claims was that an identical fusion of two specific combinations of humanoid and feline features exists on massive structures on two different worlds: Earth and Mars. Using computer enhancements of NASA's *Viking* frame 70A13 by Dr. Mark J. Carlotto (Figures 1.3 and 1.4), Hoagland produced a mirror split of the "Face" (Figures 1.5A and 1.5B). He was astonished with the results and quickly adopted the title of Martian Sphinx in reference to the "Face."

Traditionally, the Sphinx has been seen as the embodiment of man and nature. It projects man as an extension of the animal world by forming a hybrid creature that symbolizes the duality of the universe. This connection to nature speaks to the heart of humanity. If the "Face on Mars" were truly

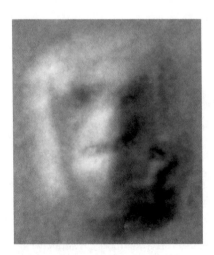

1.3 "Face on Mars" (NASA *Viking* frame 70A13).

Courtesy of NASA/JPL/Caltech

1.4 "Face on Mars" (contrast-enhanced NASA *Viking* frame 70A13).

Courtesy of Dr. Mark J. Carlotto (TASC) and Richard C. Hoagland

1.5A Hoagland's "Sphinx", left side mirrored (humanoid).

Courtesy of Richard C. Hoagland

1.5B Hoagland's "Sphinx", right side mirrored (feline).

Courtesy of Richard C. Hoagland

a Sphinx, then this simple idea of duality might hold the key to decoding the message of Cydonia. However, the answers to these questions could only be found with new, higher-resolution pictures of the Cydonia complex and the "Face."

The Lost *Mars Observer*

In a campaign spearheaded by Richard Hoagland, pressure was imposed on a reluctant NASA throughout the 1980s to re-image Cydonia with higher-resolution images during future missions. On September 25, 1992, NASA launched the *Mars Observer* spacecraft with a new, state-of-the-art camera on board capable of obtaining images at 1.4 meters per pixel. This would be a tremendous improvement over the 50 meters-per-pixel camera of the 1976 *Viking* orbiter. The exquisite camera, called the Malin Camera, was the handiwork of Arizona State University geologist Dr. Michael C. Malin, head of Space Science Systems.

NASA awarded Dr. Malin an exclusive contract to direct the camera and decide which areas of Mars to target. For the first time in NASA's history, a private individual had absolute control over its imaging system. Initially, the "Face" and its surrounding structures were not on his list. However, with intervention from the American Congress, by August 1993 NASA was finally instructed to make Cydonia a target with this new spacecraft. Soon afterward, Dr. Malin reluctantly agreed and officially announced that the Cydonia area would be targeted.

Many supporters of the artificiality hypothesis of the "Face" now anxiously awaited the new images of Cydonia. It seemed the years of controversy would soon be over. Unfortunately, on August 21, 1993 (just days after congressional intervention), NASA mysteriously lost communication with the *Mars Observer* right before it was to go into orbit around Mars. This strange event fueled the conspiracy theory of a NASA cover-up and reopened the speculations concerning the original "Face."

Demonstrators quickly gathered outside the gates of NASA and JPL, questioning the circumstances surrounding this bizarre turn of events. A NASA review board stated that there might have been a rupture in a propulsion-system line onboard the *Mars Observer* that interrupted its link with Earth. However, it was also reported that for some unknown reason the Observer's onboard computer and radio link to Earth were shut off by controllers while the fuel tanks were pressurized. When the probe's link was lost, NASA was unable to retrieve it and the mission was dead on arrival.

It would be another five years until NASA would have the opportunity to re-image the "Face" and hopefully put an end to the questions surrounding this Martian enigma.

The *Mars Global Surveyor* (the "Cat Box" Incident)

In 1996, NASA launched the long-awaited *Mars Global Surveyor* (*MGS*) spacecraft; Dr. Malin was once again at the camera's helm. The public was told that the *MGS* would thoroughly map the whole planet, including the so-called "Face," with the most detailed images ever taken of the Martian surface. The new camera could capture a dynamic range of 2048 x 4800 pixels per image, an exceptional resolution. With the capabilities of this new, high-resolution camera, expectations soared. There was no doubt that NASA would obtain a spectacular portrait of the "Face on Mars."

On April 5, 1998 the Mars Orbital Camera (MOC) aboard the *MGS* was slated to re-image the "Face," putting an end to the controversy. On April 6, 1998, Dr. Malin released a raw, distorted, low-contrast image of the "Face" to the international media (Figure 1.6) and instantly proclaimed that it was "just a pile of rocks."

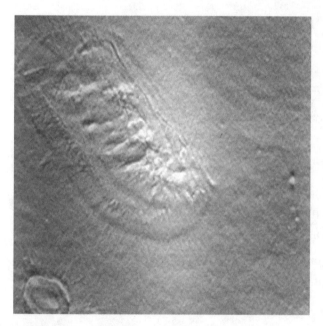

1.6 The "cat box," NASA's original press-release image of the "Face."

Taken from the Mars Orbital Camera (SP1–22003) aboard the *Mars Global Surveyor* spacecraft

Earlier in the year NASA had taken a multitude of clear, high-resolution photographs of distinct regions of Mars; however, this new image of the "Face" was so distorted and so stretched out that it was said to look more like a "sandal print or a stuffed chili pepper" by the *New York Times*.[6] The new image was of such poor quality that many critics mockingly referred it to as the "cat box" image.[7] It quickly became clear that something was very wrong with this new image of the "Face."[8]

A New Face

A few hours after releasing the distorted, low-contrast image to the international media, NASA posted a very different image (Figure 1.7A) of the "Face" on its website. Amazingly, this image had the distortion corrected and the contrast enhanced.

When the public first saw the unprocessed "cat box" image of the "Face" broadcast worldwide on April 7, 1998, they were looking for a symmetrical human face. The media and the public were unfamiliar with the analysis of raw images and, with the rapid release of this new image, they were relying heavily on the interpretation of biased "experts."

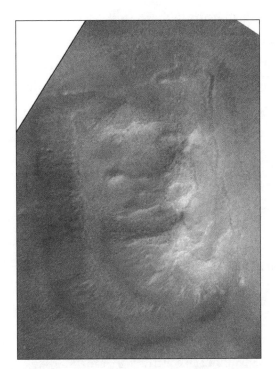

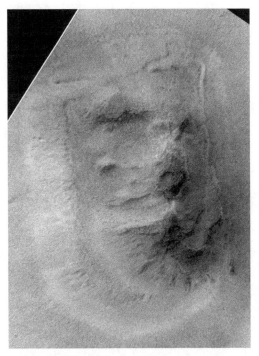

1.7A NASA's enhanced version of the 1998 "Face" from the MOC (SP–22003) (NASA's JPL-enhanced image).

Courtesy of NASA/JPL/Caltech

1.7B NASA's enhanced version of the 1998 "Face" from the MOC (SP–22003) (NASA's JPL-enhanced image, contrast reversal).

Courtesy of NASA/JPL/Caltech

Even with the rectified image posted on NASA's website, neither the media nor the public could see a familiar human face because they had no idea what they were looking at. Everyone was looking for a familiar, classic, stylized human face; however, that is not the true nature of this Martian "Face." As Richard Hoagland had most effectively demonstrated, the "Face on Mars" is not really a human face; it is an asymmetrical Sphinx-like edifice. The "Face" is like a monumental split mask that is half humanoid and half feline.

On April 12, 1998, Dr. Mark Carlotto, a leading technical engineer and developer of a new system of digital image processing at Pacific Sierra Research in Arlington, Virginia, released his first enhancement of the *MGS* "Face" on his private website. Although many in the scientific community viewed his report as inconclusive, the impact of Carlotto's detailed analysis sustained the argument that the "Face" could very well be artificial. After examining Dr. Carlotto's new enhancement, Dr. Tom Van Flandern of the Meta Research Team in Washington, D.C., a former orbital imaging specialist for the U.S. Naval Observatory, stated:

> In my considered opinion there is no longer room for reasonable doubt of the artificial origin of the face mesa and I have never concluded, "no room for reasonable doubt" about anything in my 35 year scientific career.[9]

In light of the recently condensed and enhanced version of the "Face" posted on the NASA website, a re-examination of the anomalous characteristics of this Martian Sphinx seemed to be in order. When this newly rectified version of the "Face" is split in half and each side mirrored (Figures 1.8 and 1.9), Hoagland's anticipated Sphinx-like attributes of the Martian "Face" are magnificently confirmed.

To do the split, we used an enhancement of the JPL image provided by Hoagland's Enterprise Mission website (www.enterprisemission.com 4/8/98

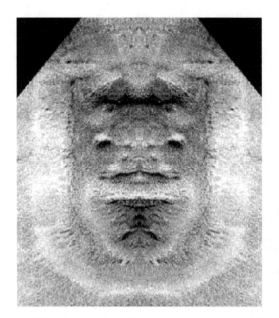 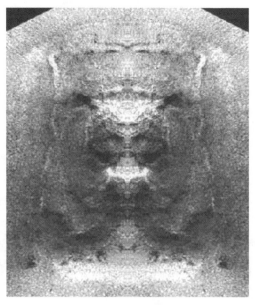

1.8 Left side of the "Face" mirrored (humanoid face, contrast reversal).

1.9 Right side of the "Face" mirrored (feline face, contrast reversal).

The image of the "Face" used for this split image is an enhancement of the JPL version, courtesy of Richard C. Hoagland

"It's a Face"). Due to the fact that the "Face" mesa was not captured from directly overhead, the feline side of the "Face" is not fully exposed. A computer enhancement, which compensates for the camera angle, provides a better sense of proportion to the image. An analytical sketch is presented in Figure 1.10 to highlight certain aspects of the mirrored "Face."

In the newly rectified image of the "Face," the most startling feature, besides the distinguishable eyes, nose, and mouth, is the presence of an elaborately styled headdress along the left side. This amazing feature runs in a straight line for over a mile. Many concerned researchers immediately saw this as further evidence supporting the artificial argument. They added this discovery to the appearance of pyramidal structures in the surrounding area and suggested that this headdress may be a link to the Egyptians.

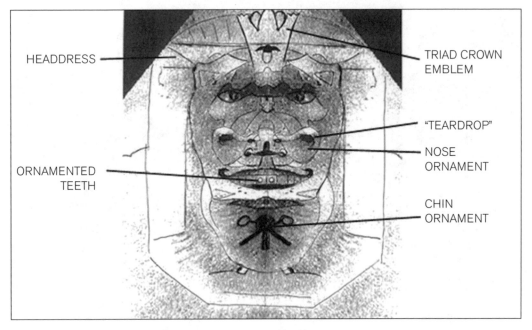

HEADDRESS

TRIAD CROWN EMBLEM

"TEARDROP"

NOSE ORNAMENT

ORNAMENTED TEETH

CHIN ORNAMENT

1.10 Analytical drawing of humanoid side of the "Face."
Drawing by George J. Haas

This interpretation is fostered by the lateral stripes (or furrows) that appear to run perpendicular to the gradual slope of the structure's base.

The combined effect of the headdress and these faint stripes that run to the ground in an orderly fashion have been interpreted by Mars researcher Mike Bara as resembling the flanged headdress of an Egyptian death mask, similar to one worn by King Tutankhamen[10] (Figure 1.12). The appearance of a second Egyptian motif is also alluded to on the forehead of the "Face" by Mike Bara on his website. An outlined object was detected at the center of the headdress that he and another researcher, Terry Kksamurai, thought looked faintly like a protruding cobra. However, when this object or mark is viewed in the mirrored version of the "Face," a very geometric, W-shaped emblem appears in the center of the forehead. Perhaps this headdress did not have a direct Egyptian connection after all.

After conducting a little research into this W-shaped emblem with various styles of cultural headdress, a match was soon discovered. Unexpectedly this Martian insignia was found to be reminiscent of the three-point leaf configuration that the ancient Maya displayed on their headdresses. In this Greenstone mask of the first century B.C. (Figure 1.11A), the Maya exhibited a three-pointed leaf emblem on their headbands to signify the crown of early kings. The basic design of this triad crown emblem was adopted by the Maya from an earlier "Mother Culture" of Mesoamerica called the Olmec.[11] The triadic, leaf-shaped glyph (Figure 1.11B) denotes the sprouting maize seed and the transformational properties of corn.[12]

It was also discovered that the Olmec, similar to the Egyptians, incorporated the lateral striped or grooved feature on their headdresses. A set of Olmec sculptures recently found in Veracruz, Mexico reveals a pair of kneeling twins wearing an Egyptian-style headdress (Figure 1.13). Surprisingly, this lateral striped effect is known among archaeologists as a typical imprint of Olmec royalty. Whatever kind of rudimentary connections the Olmec may have had with Egypt, any attempt to establish an ancient intercultural alliance between these two civilizations is strongly denied by most scholars, despite the growing evidence for it.

Another interesting attribute of Mesoamerican royalty that is incorporated in the Martian Sphinx is the use of elaborate facial ornaments. The ancient Mesoamerican people produced elaborate facial adornments out of gold, which was considered a divine substance derived from the tears of the gods. Nose ornaments were shaped in exotic designs such as geometric hummingbirds and butterflies (Figure 1.14). These ceremonial facial ornaments were so large that they sometimes covered the entire nose.

Hoagland and other researchers have been concerned about the absence of a distinct nose formation in the new MOC image. They have speculated that the nose was blown off sometime in the past by a meteorite or perhaps

1.11A Three-point crown emblem, Maya greenstone mask from Tikal (with three-point crown emblem).

Drawing by George J. Haas. (Image source: *A Forest of Kings,* by Schele and Freidel, page 115.)

1.11B Three-point crown emblem, Olmec three-pointed glyph.

Drawing by George J. Haas. (Image source: *Maya Cosmos,* by Freidel, Schele, and Parker, page 431.)

1.12 Egyptian death mask (King Tutankhamen).

Drawing by George J. Haas. (Image source: *History Unearthed,* by Woolley, 1963, page 86, fig. 80.)

1.13 Olmec twin sculpture from Loma del Zapote (El Azuzul), Veracruz.

Drawing by George J. Haas. (Image source: *Olmecs, Arqueologia Mexicana,* no. 19, National Institute of Anthropology and History and Editorial Raices , "Olmec Art," by Beatriz De La Fuente, page 37.)

by the acts of some ancient Martian war. The debris or fallout of this "major hit" distorted the nose and left an odd feature, which Hoagland called the "Tear" resting on the cheek of the "Face."[13] This remnant of the nose, later called the "Teardrop," fell within such a precise placement on the cheek that it is aligned with the center of the "City Square" in the Cydonia complex. The precise alignment and measurement of the Teardrop feature would lead one to conclude that it was actually part of an intentional design.

It is our belief that what we are actually seeing here is a large ceremonial nose ornament that obscures the nose. The Teardrop is just one part of a larger facial ornament that covers the entire nose area. This type of ornamentation over the nose is typical of the ones used by the Maya, Aztec, and most notably by the Tairona Indians of Colombia's Santa Marta mountains. The segmented nose ornament on the small four-inch Tairona pendant in Figure 1.15 resembles the bar-like design across the bridge of the nose on the humanoid side of the "Face."

1.14 Butterfly (Aztec gold nose ornament). Note that the rod went through the septum of the nose to support the adornment.

Drawing by George J. Haas. (Image source: *Aztecs: Reign of Blood & Splendor,* by Editors of Time-Life, page 138.)

1.15 Tairona warrior (gold pendant, detail). Note the segmented nose ornament and oval chin adornment.

Drawing by George J. Haas. (Image source: *Lost Empires, Living Tribes,* by National Geographic Society, page 176.)

Next, if we look at the mouth area of the humanoid side of the "Face," we may see another example of a Mesoamerican ceremonial feature. Two objects that appear to be teeth can be seen directly below the nose ornament in the mouth area. In the center of each of these front "teeth" is a dot, possibly representing a dental gemstone. This dot is similar in design to Mesoamerican decorations on the front teeth with gemstones and elaborate gold dental caps. The Maya also produced elaborate beads of jade, obsidian, or iron pyrite that were fashioned into decorative fillings and imbedded into the front teeth.[14]

Take notice of the deliberate mutilation and decoration of the upper incisors in the drawing of "Mesoamerican dentistry" from Uaxactun, Mexico (Figure 1.16). The teeth on either side of the central incisors have been filed down, making the two front teeth appear more prominent. Amazingly this same effect is displayed on the humanoid side of the "Face." Additionally, the Maya saw teeth as little skulls or kernels of corn and equated these inlaid gemstones to kernels of young green corn.[15]

We find it compelling that the humanoid side of the "Face" not only features a W-shaped emblem on its headdress that mimics the sprouted corn tri-leaf glyph, it also displays dental inlays that are similar to the Mesoamerican iconography of corn. The implications of these two corresponding features are astounding.

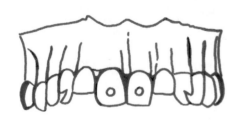

1.16 Ornamented Maya teeth. Note the circular gemstones on the front teeth.

Drawing by George J. Haas. (Image source: *The Gods and Symbols of Ancient Mexico and the Maya: An Illustrated Dictionary of Mesoamerican Religion*, by Miller and Taube, page 77.)

This new image of the "Face" shows a lot of unusual geometric ornamentation around the nose, mouth, and chin area, reminiscent of the facial adornments utilized throughout Mesoamerica. With these mask-like qualities present in the humanoid side of this Martian Face, it becomes evident that the connection between the ancient royalty of Mesoamerica and the "Face on Mars" is more than a coincidence.

Our next step in this investigation is to see if the same iconography continues on the other half of the "Face." Since Hoagland first did the split of the "Face on Mars," the feline side has always been considered a representation of a male African lion. With the new image of the "Face," the lion characteristics are even more apparent. The features of the feline "Face" when mirrored are composed of a square-shaped head with a crown, a mane, squinting eyes, an ornamented nose, an almost circular muzzle, and a mouth with fangs or teeth (Figure 1.17).

The feline's forehead is large and features a squared-off, geometric crown that extends across the top of the head. The crown also has a lot of faint decorative qualities in and around its crest that are difficult to substantiate at

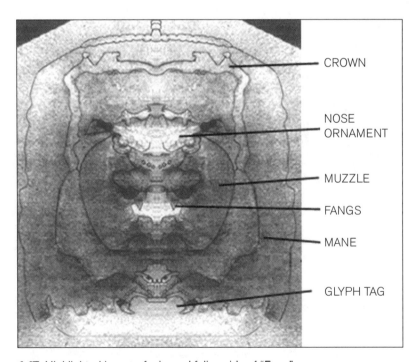

1.17 Highlighted image of mirrored feline side of "Face."

17

this point. The half portion of this crown feature was also spotted by Dr. Tom Van Flandern and referred to as the "crest" in his in-depth analysis of the MOC version of "Face."

It is known that the lion is not native to the Americas, so how can there be a connection to Mesoamerican royalty on one half of the "Face" and an African lion on the other? One answer is quite simple: The Olmec culture found in Mesoamerica was of African origin and would have been well acquainted with the "king of beasts."

There are a number of archaeologists who believe that the Olmec civilization was a result of the assimilation of both African and Asian cultures. The pre-Columbian sculptures in Figure 1.18 from Veracruz speak volumes in favor of this idea. The figure on the left has pronounced Negroid fea-

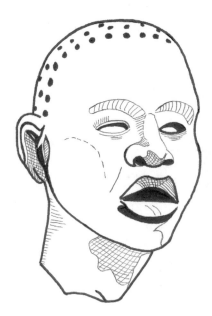

1.18A African figure in the Americas, Pre-Columbian head (Veracruz). Note the Negroid features.

Drawing by George J. Haas. (Image source: "The World's Last Mysteries," *Reader's Digest*, page 56.)

1.18B Asian figure in the Americas, Costumed figure of a woman (Veracruz, seventh century A.D.). Note the Asian-style dress.

Drawing by George J. Haas. (Image source: "Mysteries of the Ancient Americans," *Reader's Digest*, page 211.)

tures while the figure on the right looks very much like an Asian kabuki dancer.

The "King of Beasts" in Mesoamerica?

Mesoamerica had its own "lion" in the form of the jaguar. It would be reasonable to assume that the Olmec would have regarded the jaguar as the Mesoamerican equivalent of the lion. Since the jaguar does not have a mane, it would also be reasonable to assume that the indigenous Mesoamericans would have regarded the lion with its full mane (Figure 1.19) as a bearded jaguar. In the National Museum of Anthropology in Mexico City sits a large Aztec reliquary that is carved in the shape of a lion. However, because of its geographical and cultural location, it is officially labeled a jaguar (Figures 1.20 and 1.21).[16] This amazing sculpture, which weighs over six tons, was unearthed at Templo Mayor in Mexico City in 1790.[17]

The most intriguing characteristics of this so-called jaguar sculpture are that it has no spots—jaguars do—and it has a mane—jaguars do not. The

1.19 Male African lion (a bearded jaguar?).

1.20 Aztec jaguar reliquary (side view). Note the mane and the absence of jaguar spots.

Drawing by George J. Haas. (Image source: *Myths of the World: Gods of the Inca, Aztec and Maya*, by Roberts, page 67.)

1.21 Aztec jaguar reliquary (front view). Notice the shape of the partial mane and its resemblance to the mane on the Mars feline.

Drawing by George J. Haas. (Image source: *The Mighty Aztecs,* by Stuart, photos by Mark Godfrey, © The National Geographic Society, 1981, page 194.)

1.22 Bearded jaguar god (Maya glyph).

Drawing by George J. Haas. (Image source: *The Code of Kings,* by Schele and Mathews, page 409.)

mythology of the jaguar god is associated with the Maya god archaeologists call GIII, who is human in aspect and has jaguar features. The most provocative attribute of this anthropomorphic jaguar god is that he is also presented with a beard and is referred to as the bearded jaguar (Figure 1.22).

The First Temple (a Face on the Horizon)

On the face of a two-tiered ancient Mayan pyramid that was intentionally buried around 50 B.C., some fascinating sculptures appear. Archaeologists call this pyramid (Figure 1.23), located on a small peninsula at Cerros in what is now Belize, the First Temple (Structure 5C–2nd). Occasionally, the Maya practiced a strange ritual of "urban renewal" in which they would actually bury temples or entire villages, and then build new structures over them.[18] Because of this practice, the elaborately carved structures on the temple are in relatively good shape.

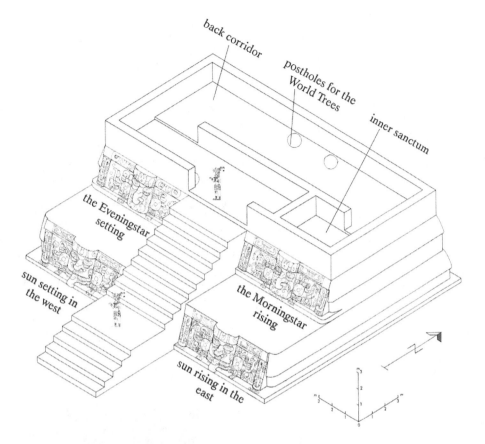

1.23 The First Temple at Cerros (reconstructed projection).

Drawing by Linda Schele, © David Schele. (Courtesy of Foundation for the Advancement of Mesoamerican Studies, Inc., www.famsi.org)

On the face of the temple, four decorated panels with large plaster-covered masks flank a central stairway. The top two masks represent the planet Venus—as the morning star on the eastern side and the evening star on the western side. The lower masks represent the Jaguar Sun god: as the rising sun in the east and the setting sun in the west. Venus and the Jaguar Sun are also representations of the Maya twin gods known as First Lord (Venus) and First Jaguar (Jaguar Sun).[19] First Lord was the god associated with resurrection, while First Jaguar was the god associated with the Underworld. This temple pyramid is seen as a cosmological diagram that joins the Heavens with the Underworld.

In describing the First Temple at Cerros in their book *A Forest of Kings: The Untold Story of the Ancient Maya,* Mayanist Linda Schele and archaeologist David Freidel state:

> The first temple was in the center of the vertical axis that penetrated the earth and pierced the sky, linking the supernatural and natural world into a whole. This plan set the temple between the land and the sea on the horizontal axis and between the Heavens and the Underworld on the vertical axis.[20]

The panels "were designed to be read as symbolic statements about the nature of the kingship and its relationship to the cosmos."[21] Within the design of the First Temple at Cerros, with its symbolic pairing of the Sun and Venus as the cosmic twins (Figures 1.24 and 1.25), we discover the

1.24 Temple mask–Venus (Cerros). Note the triad crown emblem and the odd facial ornaments around the nose, mouth, and chin area.

Drawing by Linda Schele, © David Schele. (Courtesy of Foundation for the Advancement of Mesoamerican Studies, Inc., www.famsi.org.)

1.25 Temple mask–Jaguar Sun (Cerros). Note the snarling aspect of the snout and the glyph tag under the neck.

Drawing by Linda Schele, © David Schele. (Courtesy of Foundation for the Advancement of Mesoamerican Studies, Inc., www.famsi.org.)

same fusion of human (Venus) and feline (Jaguar Sun) aspects that we find in the "Face on Mars."

Figure 1.26 demonstrates the amazing similarities between the morning star mask on the Mayan temple and the humanoid "Face on Mars." Both present the same triad crown emblem on the headband and similar facial ornaments in the nose and chin area, including the Teardrop feature. When the feline side of the "Face" is compared to the Jaguar Sun mask on the lower panel of the temple, it takes on a very different persona. It is apparent that both have the same square-shaped face and exhibit a crest-like crown on and above the forehead (Figure 1.27). Notice the snarling aspect of both faces and the curvature of their mouths.

The Jaguar Sun god is portrayed as both the Bearded Jaguar and the Sun, as the comparison in Figure 1.28 clearly illustrates. If you look at just the cir-

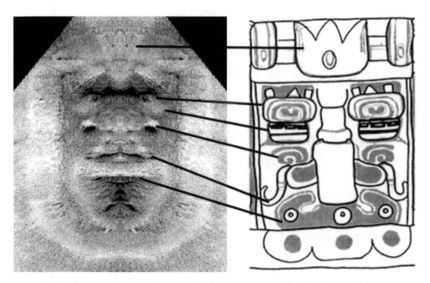

1.26 Comparison of mirrored humanoid side of the "Face on Mars" with Venus temple mask.

LEFT, Mirrored humanoid side of "Face on Mars."

RIGHT, Venus mask (from the upper panel of the First Temple at Cerros). Note the facial ornaments and W-shaped emblem on forehead.

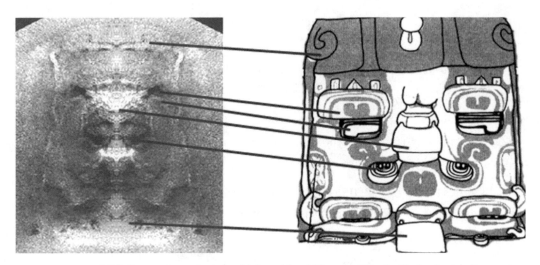

1.27 Comparison of mirrored feline side of "Face on Mars" with the Jaguar Sun god temple mask. LEFT, Mirrored feline side of "Face on Mars." RIGHT, Jaguar Sun (from the lower panel of the First Temple at Cerros).

cular muzzle and the flaring face of Jaguar Sun, a comparison to the common diagrammatic representation of the Sun is easily made. Emerging like a totem under the chin and along the neck line of the Jaguar Sun mask is a portion of a second mask. This partial mask (Figure 1.29) portrays a long-snouted creature with a glyph tag in the center symbolizing the horizon.[22] By combining the Cerros jaguar mask with this glyph tag, its meaning is amplified and the panel reads "Jaguar Sun at the horizon." This new bit of cosmological information is intriguing.

In his book *The Monuments of Mars,* Hoagland talks about the importance of the placement of the "Face" at Cydonia and the horizon point, with the "Cliff" as a backdrop.[23] The Cliff is a structure in Cydonia that Hoagland feels was constructed to form a horizontal ridge line as a perfect backdrop for viewing the "Face" from the "City Square."[24] He goes on to discuss the Egyptian god Horus (whose real Egyptian hieroglyphic designation was Heru) as the god of the setting and rising Sun. He found that the

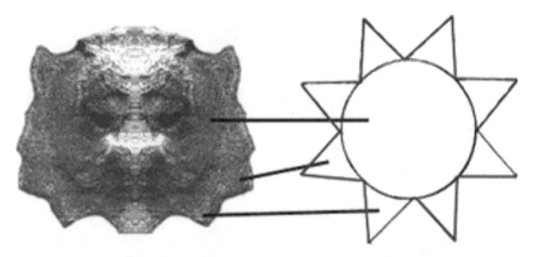

1.28 Comparison of the muzzle of the mirrored Jaguar Sun face with a common diagrammatic representation of the sun.

translation for the term "Horus of the Horizon" actually meant "Face on the Horizon."[25] He states:

> ... Heru can equally apply to "the sun on the horizon," a race "between gods and men," a "face" (on the horizon?) ... and the "king as the representative of the Sun-god.... [26]

Just as the lion is acknowledged as a symbol of the Sun in Egypt, in Mesoamerican culture it is the jaguar or rather the bearded jaguar that represented this celestial body. Schele and Freidel confirm this relationship:

> ... the whole message of the [Cerros] temple comes into focus with these Sun Jaguars. Since this building faces to the south, a person gazing at its colorful facade would see the Sun in its jaguar aspect "emerging" from the sea on the eastern side of the building and "setting" into the sea on the western side.[27]

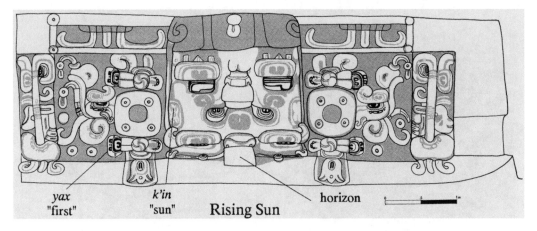

yax　　　　*k'in*　　　　　　　　　　horizon
"first"　　　　"sun"　　**Rising Sun**

1.29 Jaguar Sun god (panel from bottom right side of the First Temple at Cerros). Note the glyph tag in the lower center, indicating "Horizon."

Drawing by Linda Schele, © David Schele. (Courtesy of Foundation for the Advancement of Mesoamerican Studies, Inc., www.famsi.org.)

The identical symbolism with Egypt is unmistakable. The lion represented the king in ancient Egypt, the jaguar (or bearded jaguar) represented the king to the Maya. Osiris, as the setting sun, represented the dying or dead king and was known as the "god of the Underworld." To the Maya, the Jaguar Sun was the "god of the Underworld." The rebirth of the dead king in the form of his son in Egypt was Horus, "the god of the rising sun." To the Maya it was First Lord or Venus as the morning star that represented rebirth or resurrection as it led the new sun above the horizon. The pyramid texts in Egypt also refer to Horus as the morning star.

Were the architects on Mars as consumed with the movements of their own morning star, Earth, as the Maya and Egyptians were with theirs? According to Hoagland, the precise placement of the "Face on Mars" would cause an amazing event to happen on the very day of the Martian summer solstice. Hoagland explains it like this:

> ... if one of those "hypothetical Martians" had stood in the center of "my" City in the "City Square," they would have seen the Earth rise

brilliantly in the dawn. And just a few moments later the Sun would have "magically" appeared ... rising out of the mouth of the god-like figure. [However] ... the last time this alignment would have worked was half a million years ago.[28]

From what we have observed in the two temple masks at Cerros in reference to a Mesoamerican model of design, it is no longer a viable argument to assert that the "Face on Mars" is nothing more than a "pile of rocks." Although the "Face" was almost entirely eradicated and dismissed from our consciousness as a "trick of light and shadow," it is now clear that its facade is no optical illusion. Like the temple masks at Cerros, the Martian Sphinx may have once been covered up, but now its masked secrets are beginning to be resurrected.

Most notable is the fact that NASA may have been aware of the resemblance between the Martian visage and the mask at Cerros years before we were. We find it very interesting that world-acclaimed Mayanist and scholar Linda Schele was invited to speak before NASA in 1995 at a seminar entitled "The Universe: Now and Beyond."[29] Considering her intimate knowledge of the Maya cosmology and iconography, what did NASA wish to learn? Throughout her lecture, Schele chronicled the vast knowledge of the Maya and its broad accessibility to the masses, which is encoded within their lifestyles and creation myths.

In retrospect, it makes one wonder, what did NASA already know when they took this new image of the "Face" in 1998 and how long have they known it? Incredibly, the First Temple at Cerros is not our only bit of Mesoamerican evidence. It is only the first link in a long chain of terrestrial connections between the "Face on Mars" and its lost kinship to the ancient Olmec and Maya. Not far away and just below the Face lies another major link.

*Authors' note: We will no longer use quotation marks around the word **Face*** *to designate the Face on Mars, as in our opinion there is no longer any doubt about its artificial design and its split-faced nature.*

Notes

1. Randolfo Rafael Pozos, *The Face on Mars, Evidence for a Lost Civilization?* (Chicago: Chicago Review Press, 1986), 7–8.

2. Vincent DiPietro and Greg Molenaar, *Unusual Martian Surface Features,* 3rd ed. (Maryland: Glen Dale, 1982), 38.

3. Pozos, ibid. DiPietro and Molenaar found additional structures that they felt revealed intelligent design, most notably pyramids. One polygonal structure in particular, known as the D&M pyramid was named in their honor.

4. Vincent DiPietro, "Report of Findings of the Face on Mars in the Cydonia Region," 1 July 1996. http://www.lauralee.com/face.htm. DiPietro discovered ten images depicting the "Face." Six of the images are low-resolution and two are medium-resolution. The two that are high-resolution were reexamined using digital image-enhancement techniques. The following is a NASA archival list with frame numbers of the complete set of "Face" images from the *Viking* orbiter: 35A72, 70A13, 56A25, 561A27, 673B54, 673B56, 753A33, 753A34, 814A07, and 257S69.

5. A Soviet geologist, Dr. Vladimir Avinsky, was actually the first to refer to the "Face on Mars" as a Sphinx, in the August 1984 issue of *Soviet Life* magazine.

6. "New Mars Photos Cast Doubt on Speculation on a 'Face,'" *New York Times,* April 7, 1998, A24.

7. The term "cat box" was coined by radio host Art Bell ("Coast to Coast") on April 6, 1998 and adopted by many analysts, after a caller's comparison of the *MGS* Face image to a cat's litter box.

8. NASA decided to extend the down-track capabilities of the camera, at the expense of the cross-track and image resolution, in order to ensure capturing the "Face" mesa. Thus, the resolution was reduced from 2.1 meters per pixel to 4.3. According to other experts, the problem in cap-

turing the "Face" mesa was not in the down-track capabilities but in the cross-track, evidenced by the fact that the mesa was in the center of the down-track range. Before the *MGS* image was released to the public it was processed through a "high-pass filter," according to JPL's own image log. This process suppresses detail and is normally utilized with line drawings and high-contrast black-and-white pictures. A sound reason to use such a filter has never been successfully offered by NASA or JPL. The new image was found to have only 42 shades of gray, while a normal MOC image was capable of 256. The distortions in the raw image can be confirmed by simply examining the elongated oval shape of the crater just below and to the lower left of the "Face" mesa. In the original 1976 *Viking* image of the "Face," it is clear that this crater is perfectly round! See Richard C. Hoagland, "Honey I Shrunk the Face," The Enterprise Mission, April 14, 1998. http://www.enterprisemission.com/

9. Tom Van Flandern, Meta Research, April 1998. http://www.metare-search.org/

10. Mike Bara, "New Mars Face Image Analysis and Comment, Part III," The Lunar Anomalies, 1999. http://www.lunaranomalies.com/

11. Linda Schele and David Freidel, *A Forest of Kings: The Untold Story of the Ancient Maya* (New York: Quill, 1990), 115.

12. Linda Schele, David Freidel, and Joy Parker, *Maya Cosmos:Three Thousand Years on the Shaman's Path* (New York: Quill, 1993), 431.

13. Richard C. Hoagland, *The Monuments of Mars: A City on the Edge of Forever,* 4th ed. (Berkeley: North Atlantic Books, 1992), 22.

14. J. Eric S. Thompson, *The Rise and Fall of Maya Civilization* (Norman, Okla.: University of Oklahoma Press, 1966), 214.

15. Karn Bassie-Sweet, "Corn Deities and the Complementary Male/Female Principle," Mesoweb, September 7, 2000. http://www.mesoweb.com/features/ bassie/corn/index.html

16. Occasionally archaeologists discover sculptures of animals in Mesoamerica that are not indigenous to the Americas. In 1995 author and historian Zecharia Sitchin visited the Anthropological Museum in Jalapa, Mexico and photographed a vessel in the shape of a small clay ele-

phant dated to the Olmec period. See Zecharia Sitchin, *The Earth Chronicles Expeditions: Journeys to the Mythical Past* (Rochester, Vermont: Bear & Company, 2004), 71, 72, plate 20.

17. Timothy R. Roberts, *Myths of the World: Gods of the Maya, Aztec and Incas* (New York: Metro Books, 1996), 67.

18. Linda Schele and David Freidel, op. cit., 103.

19. Ibid., 117.

20. Ibid., 105.

21. Ibid., 113.

22. Ibid., 113.

23. Richard C. Hoagland, op. cit., 76.

24. Similar architectural backdrops were employed by the Olmec in the construction of the La Venta complex in Mexico. Precisely placed mounds formed backdrops that created a ritual stage where cosmic ceremonies of creation were reenacted. Richard A. Diehl, *The Olmecs: America's First Civilization* (London: Thames & Hudson, 2004), 61, 64.

25. Richard C. Hoagland, op. cit., 287.

26. Ibid., 288.

27. Linda Schele and David Freidel, op. cit., 114.

28. Richard C. Hoagland, op. cit., 64.

29. "NASA Administrator's Third Seminar Series Scheduled," *NASA News* March 13, 1995. http://www.aero.com/news/nasa_press/n950308d.txt. A video of Linda Schele's NASA presentation "The Universe: Now and Beyond" was acquired by George Haas from Gillett Griffin of Princeton University in the spring of 2002.

CHAPTER 2

Cydonia, A Mosaic of Faces

The First Three Swaths

In the spring of 1998, the Mars Orbital Camera on board the *Mars Global Surveyor* spacecraft imaged portions of the Cydonia region in three separate swaths. Each swath was acquired as a long, narrow, north-to-south strip several kilometers wide by several tens of kilo-meters long. The first swath was taken on April 6, 1998, and featured the famous Face on Mars plus a second geoglyphic mesa called the "Animal" (Figure 2.1). The first swath began far above the Face and continued down to the edge of the D&M pyramid from a dis-tance of 444 km (275 miles) above the sur-face of Mars. The second strip, taken on April 14, 1998, focused on a mountainous area including a section that was deemed to rep-resent the outskirts of the "City" (Figure 2.2). The third strip, taken on April 23, 1998, cap-tured the "City Center" (Figure 2.3). NASA posted these photographs on its web site.

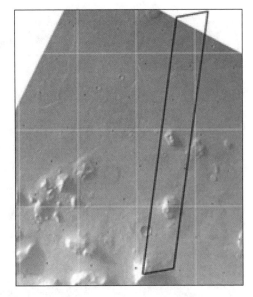

2.1 Location of first Cydonia swath (SP1–22003) (4.42 km x 41.5 km).

Courtesy of NASA/JPL/Caltech

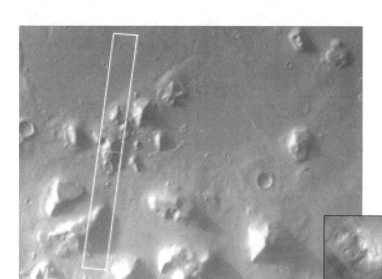

2.2 Location of second Cydonia swath (SP1–23903) (2.5 km x 24 km).

Courtesy of NASA/JPL/Caltech

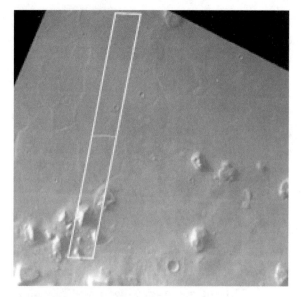

2.3 Location of third Cydonia swath (SP1–25803) (3.5 km x 33.2 km).

Courtesy of NASA/JPL/Caltech

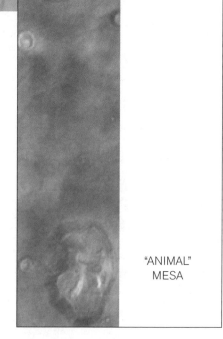

THE "FACE"

"ANIMAL" MESA

2.4 Context image: location of Animal mesa (with respect to Face–MOC SP1–22003).

The Animal

It became apparent to us, upon examining this first swath on the NASA web site, that the Face was not alone. Below the face at the center of the swath was another anomalous mesa. This second mesa, roughly the same size as the Face, also had a bifurcated geoglyphic appearance (Figure 2.4). Although this geoglyph was easy to find, we felt it was odd that nobody had acknowledged it before. It seemed that the construction of this mesa went far beyond a series of random geologic processes.

Upon conducting a quick search on the web, we found a discussion of the mesa on Dr. Tom Van Flandern's website. Van Flandern casually referred to this mesa as the Animal. We agree that the mesa does look like an animal's face, but we saw two separate faces.

We inverted the mesa to what we thought was the correct orientation for viewing by rotating the image 180 degrees (Figure 2.5). Some unusual, noncontinuous markings were noted across the top of the mesa. The eastern and western sides of the structure appeared to collide, suggesting a demarcation running down its middle. And while an oblique camera-angle shot had captured the Face, the second mesa had been captured from almost directly overhead, so its duality could be readily seen. Inspired by Hoagland's mirror flip of the original Face, we performed a mirror flip on this formation as well. What we observed was a set of oddly matched creatures. The left side looks like the face of a monkey (Figure 2.9), while the right side resembles a rabbit (Figure 2.6).

On the top half of Figure 2.6 there appears to be a geologic uplifting or rampart in the shape of a

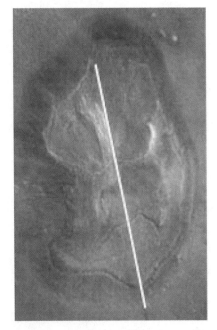

2.5 The Animal mesa (inverted) (with diagonal demarcation added).

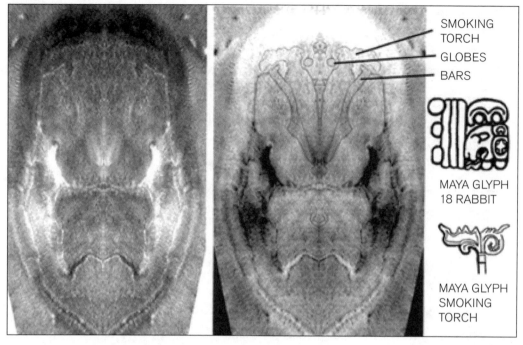

SMOKING TORCH
GLOBES
BARS

MAYA GLYPH 18 RABBIT

MAYA GLYPH SMOKING TORCH

2.6 Martian rabbit.
LEFT, Mirrored image of rabbit (18 Rabbit?).
RIGHT, Contrast inversion highlighted (with Maya glyphs added).

rabbit's ears. Between the ears are a set of bars, globes, and what is believed to be a "smoking torch" symbol that forms a crown. Farther down at the center of the face is a faint wedge shape that outlines the cavity of a small eye. Below the eye socket is a raised cheekbone and a muzzle that extends down to a clearly defined ridge line. This swerving ridge forms and shapes the outline of a mouth and jowls.

The head-on presentation of the mirrored rabbit's face is that of a crowned animal head that appears to be displayed like a trophy on a plaque. The spiked crown formation between the ears brings to mind the dot-and-bar method of counting used by the Maya. In the Mayan counting system, each bar equals the number 5 and a dot equals 1. Could these bar-like spikes and spherical globes be identifying this image as that of the Maya king known as 18 Rabbit?

The glyph for 18 Rabbit is a profile of a rabbit head with three bar-and-dot symbols placed in front of his forehead. The smoking torch on top of the rabbit head was a symbol the Maya used to denote a supernatural being. It is believed the symbol is being used to depict 18 Rabbit as a Uay (or Way). A Uay was a divine manifestation of a human being, or a god in its companion-animal form. For the ancient Maya, "the Uay dwelled in the shadowy region between individual human identity and the deeper realm of the gods and demons."[1]

According to the mythology of the Maya, the rabbit is also a symbol of the Earth and as such is named 1 Rabbit.[2] He is also the consort of the Moon goddess, who is called Xt'actani.[3] On a unique ceramic sculpture produced by the Maya, the couple can be seen embracing, demonstrating the relationship between the Earth (1 Rabbit) and the moon (Xt'actani) (Figure 2.7). Notice the stylized treatment of the rabbit's head. When this idealized Mesoamerican image of the rabbit is compared to the Martian rabbit, the similarities are quite astounding.

Upon closer examination of this finely crafted sculpture of the 1 Rabbit and the Moon goddess, we noticed the presence of a large patch of fur that hangs like a flange along the jaw line of the rabbit. These radiant lines and curls form a furry collar like those worn by kings. Similar radiant lines can be seen along the border or neck line of the Martian rabbit. Dr. Tom Van Flandern originally noticed this anomalous feature on the original

2.7 1 Rabbit with the Moon goddess (Maya ceramic).

Drawing by George J. Haas. (Image source: *Hidden Faces of The Maya*, by Schele.)

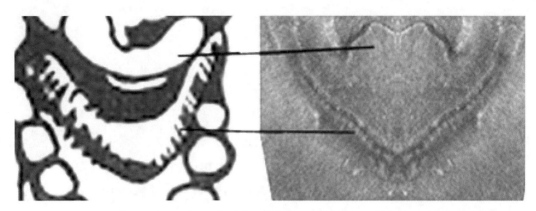

2.8 Comparison of mouth and neck features of Maya and Martian rabbits.
LEFT, Mouth and neck area of Maya god 1 Rabbit.
RIGHT, Mouth and neck area of mirrored Martian rabbit.

mesa in April 1998 and referred to it as an "aura." Could this aura of regularly spaced white lines (Figure 2.8) represent fur radiating around the neck of the Martian rabbit?

The companion face to the Martian rabbit is also an odd-looking creature; it resembles a primate such as a monkey or an orangutan (Figure 2.9). The odd shape of the jaw is similar to that of a mature male orangutan. At the center of these broad jowls is a small, bow-shaped mouth. The shape of the mouth is formed by an extension of a fluctuating ridge line that starts on the left half of the mesa. If you follow the length of the full ridge you will notice that it transversely forms a common mouth line for both faces. Above the mouth line is an ape-like nostril and a short muzzle. Notice the statue within the nose of the primate and how its wings conform to the arch of the nostrils (Figure 2.10). This compound feature, which displays the idea of figures within figures, is a common Mesoamerican design technique. Could this be the representation of another Maya king or god?

Just above the nose two small eyes rest at the sides of what we believe to be another smoking-torch symbol extending from the bridge of the muz-

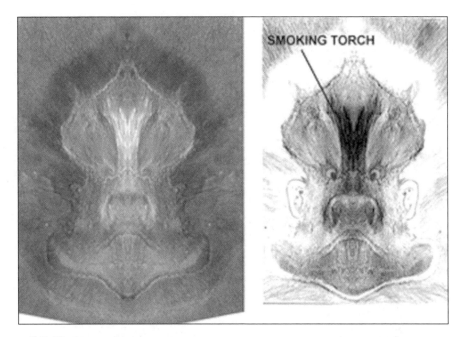

SMOKING TORCH

2.9 Martian monkey.
LEFT, Mirrored image of Smoke Monkey.
RIGHT, Contrast inversion (with facial features highlighted).

zle up through the forehead. A smoking-torch symbol was also attributed to the Maya king Smoke Monkey, the successor to the most famous Maya king of Copan—18 Rabbit.

"Thirty-nine days after the defeat of 18 Rabbit, on a day close to the maximum elongation of Venus as morning star, a new king named Smoke Monkey acceded to the throne."[4]

It's obvious in the quote above that some kind of cosmological reference between 18 Rabbit and Smoke Monkey is encoded in this text. Is this cosmological symbolism harking back to the duality

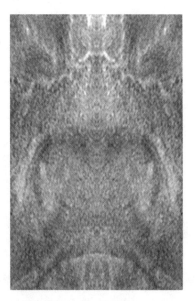

2.10 Detail of muzzle area of the monkey side of the Animal mesa. Note the winged figure within the muzzle area of the Smoke Monkey image.

2.11A Maya Rabbit scribe (detail of ceramic cylinder vase).
Drawing by George J. Haas. (Image source: The Princeton Vase MS1404.)

of the original Face and its connection with the masks found at the First Temple at Cerros?

The timing of Smoke Monkey's accession to the throne is very interesting, considering what we have learned about the original Face in Chapter 1. It is also amazing that these depictions of two Maya kings, who were contemporary rivals in Mayan history, are found side by side (or face to face) on a single geoglyphic structure on Mars! Traditionally, the rabbit and the monkey have similar symbolic associations. They are both seen as the "Trickster" and are fertility symbols associated with the Moon. The rabbit and the monkey are also both portrayed as scribes on Maya pottery (Figures 2.11A and 2.11B).

2.11B Maya Monkey scribe (detail of polychrome vase).
Drawing by George J. Haas. (Image source: Justin Kerr K5744.)

The *Viking* Mosaic

After discovering the second geoglyphic split face in the first MOC strip, we began to re-examine the early *Viking* orbiter data.[5] We used readily available images of the Cydonia area that had been previously enhanced by other researchers. The source image we used was a *Viking* mosaic of the Cydonia complex produced by Dr. Mark Carlotto.

At the time, Dr. Carlotto's enhancements were the best available images of this vast Martian complex to be found, and we felt they warranted a closer look. After contacting him about our interest in his enhancements, we were graciously supplied with a digital copy of his orthographically rectified photomosaic of the entire Cydonia complex (Figure 2.12).

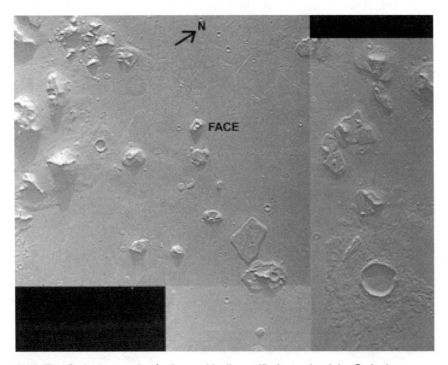

2.12 The Cydonia complex (orthographically rectified mosaic of the Cydonia-area *Viking* photos).

Courtesy of Dr. Mark J. Carlotto

The Third Face

With fresh insight into the Mesoamerican model of design, prevalent in the landforms at Cydonia, we found our perception of these split-faced structures was now clearer. We began examining these anomalies, starting with the third split-faced structure that we discovered, which is situated just below and to the east of the original Face (Figure 2.13).

At first the structure, which is well over a mile in length, was thought to present a comical pairing of a reptilian head and a smiling feline. One half of a reptilian face can be seen on the left side of the mesa and one half of a playful feline face is on the right (Figure 2.14).

The reptilian aspect of this Martian geoglyph resembles a broad-faced turtle's skull. When mirrored, the framed shape of the eye socket forms two large circular pockets that echo the outline of eye goggles. Between the goggles is a beak-like nose bridge on which the goggles rest. The trian-gular shape of the nasal area broadens as it takes on the form of a pushed-

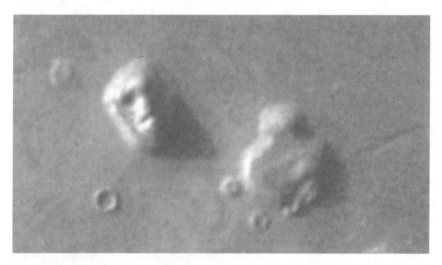

2.13 Context image for location of Third Face mesa (in relation to original Face on Mars).

Courtesy of NASA

up snout. Below the nose is a zigzag-shaped ridge line. This traversing line delineates the edge of the lips and lower jaw, forming the appearance of a rippling mouth. Affixed to the left side of this geoglyphic structure is a pair of bell-shaped craters.

After a great deal of debate we took another look at the bell formation on the Third Face. Was it an intended part of the image, or were these just random craters? If these were indeed craters the double impact had no effect on the sur-

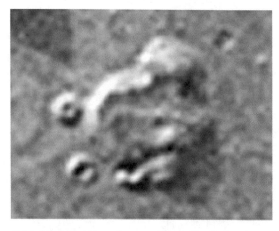

2.14 The Third Face (cropped from Dr. Carlotto's *Viking* mosaic).

Courtesy of Dr. Mark J. Carlotto

rounding terrain. After closer examination, we noticed that the zigzag movement of the mouth forms a rippling serpentine configuration that extends right through the mouth area and toward the bell-shaped crater. Is the curvaceous ridge compliant with the modeled body of a snake? Is the lower bell feature actually the head of a snake whose body is clenched in the mouth of a turtle-like creature? If so, we wondered if this half of the Third Face could be the Maya Witz Monster, with a snake in its mouth.

To the Maya, the Witz Monster is the symbol of the living mountain (Figure 2.15). *Witz* is the Mayan word for mountain or hill.[6] When a snake emerging from its mouth is added to the design of the mask, the Witz Monster takes on the personification of the Snake Mountain (Figure 2.16). Notice how the head of the snake intertwines throughout the ear spool of the Snake Mountain temple mask. To the Maya, the fabled Snake Mountain is the place where civilization was invented. In Mesoamerican myths, the Maize god dies and is reborn through a crack in the top of a mountain or through a crack on the back of the Cosmic Turtle.[7] When compared, the

2.15 Maya Witz Monster mask (from Temple 22 at Copan).

Drawing by George J. Haas. (Image source: *A Forest of Kings*, by Schele and Freidel, page 73.)

left side of the third face has a compelling resemblance to the Maya glyph for turtle (Figure 2.17). Notice that the profiled glyph of the turtle has a beaked nose and a large hooded eye.[8]

If the left side of this mesa is the Martian equivalent to the fabled Snake Mountain, or Witz Monster, there should be key facial similarities. The following example of the Third Face is a mirrored image of the left side of Carlotto's enhancement, which chronicles the similarities between this two-faced structure and Mesoamerican iconography (Figures 2.18A and 2.18B). Both masks have the same balloon-shaped crest over the forehead. The large scrolled horns that project from the balloon crest in the temple mask are variant glyphs that sig-

2.16 Maya Snake Mountain mask (from structure 5D-33-2nd at Tikal).

Note the snake that twines through the lower ear spool or porthole.

Drawing by Linda Schele, © David Schele. (Courtesy of Foundation for the Advancement of Mesoamerican Studies, Inc., www.famsi.org.)

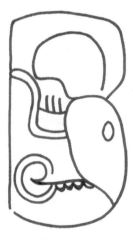

2.17 Maya glyph for turtle. Note the beak-like nose and large eye.

Drawing by George J. Haas. (Image source: *Maya Cosmos*, by Freidel, Schele, and Parker, page 66.)

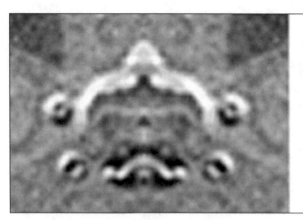 

2.18A Snake Mountain comparison. Martian Snake Mountain mask (mirrored).

From Dr. Mark Carlotto's Viking mosaic

2.18B Snake Mountain comparison. Maya Snake Mountain mask.

Drawing by Linda Schele, © David Schele. (Courtesy of Foundation for the Advancement of Mesoamerican Studies, Inc., www.famsi.org.)

nify blood or maize in Mayan iconography. This feature has a pebbled effect in both images.

The large swirls at the sides of the eyes on the temple mask are in a slightly lower position on the Mars mesa and also have a pebbled effect. The same broad, beaked nose, which resembles the beak of a turtle, can be identified in both masks. The most convincing feature is the common serpentine ridge line that starts in the mouth and curls around and through the spool at the side. On the Mars mesa it protrudes downward from the mouth in an upside-down V and bends around so the snake head eventually appears out of the loop or portal (lower crater) at the side. Even with the poor *Viking* image of this mesa, we believe we have another terrestrial match. According to the Maya creation myth, when the Witz Monster is in the form of the Snake Mountain, there is a ball court at its base. In the ball court, the loser of the game is sacrificed by decapitation as a re-enactment of the death and rebirth of the Maize god.[9]

The oblique camera angle of this *Viking* image does not allow a very clear view of the entire right side of the mesa, and the angle of the sun leaves most of the features on the right side in shadow. Fortunately Carlotto's enhancement of this image brings a great deal of detail to light. When the right half of his enhancement is mirrored (Figure 2.19), the whimsical demeanor of a feline face can be identified.

Notice the extended grin, and the muzzle that supports the M-shaped nose. Just above and to the side of the muzzle is a large, dark, oval shape that forms the recess of the eye socket. Due to the shadow, no eye formation can be established; however, the eye socket is darker in the center. Next, notice the round hillock configuration that mimics the extension of a tight, muscular cheekbone. Its elevation draws up the mouth line to form an expanded grin. The mold of this tight cheekbone, including the grin, is indicative of the facial characteristics expressed by the Mayan god of Decapitation.

On the Gulf Coast of Mexico, at a site called Veracruz, a small ball-court

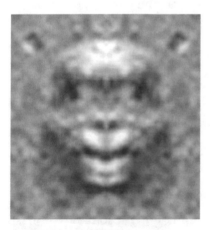

2.19 Martian Decapitation god (mirrored). Note the feline features, the M-shaped nose, and the extended grin.

From Dr. Mark Carlotto's Viking mosaic

2.20 Decapitation god (Maya ball-court hacha, A.D. 550). Note the feline features, the M-shaped nose, and the extended grin.

Drawing by George J. Haas. (Image source: *Sport of Life and Death*, by Wittington, page 59.)

hacha was found that reveals a striking similarity to the Martian feline face. (Figure 2.20). This fine example of a greenstone hacha depicts a severed head that clearly displays the anthropomorphic features of the human-feline aspects of the Decapitation god. The Olmec and Maya produced these thin stone faces that, because of their carved tin edges, resemble an axe head. These axe-like objects were worn as an appendage to the ballplayers' yoke during the ball game, as surrogate trophy heads.[10]

Notice the skull-like quality of the hacha of the Decapitation god: the exaggerated grin and the tusk-like fangs that curl around the cheek area and the small, M-shaped nose (Figure 2.19).

The pairing of these two related faces, the Snake Mountain mask and the mask of the Decapitation god, is significant because both characters play prominent roles in the Maya creation myth. The Snake Mountain mask (on the left) represents the idea of the ball court and the point of creation, while the Decapitation god mask (on the right) alludes to the ritual ball game of life and death played by both the Maya and the Olmec. This particular mesa on Mars is another link to the analog between two worlds that use a common lexicon—split faces.

Although the data provided in Carlotto's enhancement of this Third Face mesa suggest a strong facial similarity to Mayan iconography, further high-resolution pictures are still needed to substantiate this terrestrial connection. Hopefully, the MOC will target this mesa in the future.[11]

The Winged Head

The next type of facial structure recognized was individual half-faced profiles. A common attribute of these facial structures is that they abruptly end at the centerline of each face, creating an offset profile. These faces are not split faces because at the point where the half-face abruptly ends, a featureless level plane begins.

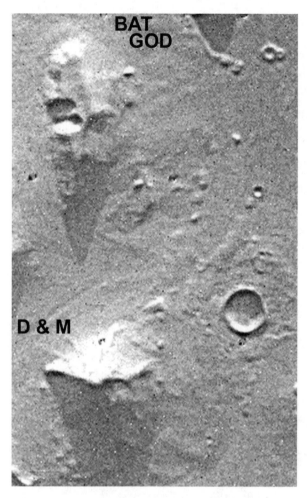

2.21 Context image of the half-faced geoglyph (with winged headdress).

From Dr. Mark Carlotto's Viking mosaic

The first half-faced geoglyph that we discovered was on a mesa just to the left of a structure known as the D&M pyramid (Figure 2.21). This anomalous mesa, more than a mile and a half in length, is composed of a half image of a human-like head with a bat wing protruding from its side. When this structure is mirrored, a second face can be seen directly below the image of the human head with the bat-wing headdress (Figure 2.22). If you look very closely, you can see the totemic head of a winged skull that begins directly under the chin of the main image in Figure 2.24. Is this bat-winged structure at Cydonia the Martian equivalent to the Bat god of the Maya? Is this another example of an ancient archetype that we share with Mars?

On the face of a 3,000-year-old jade pendant, a portrait of an early Olmec were-jaguar is depicted with two wing-like flanges flanking its head (Figure 2.23). Notice the cleft in the forehead and the curved wing-like shape of the flanged headdress. A similar winged motif is found in Mayan iconography. The Maya Bat god, called Camozotz, is represented as half man, half bat, with bat wings protruding off the sides of his head. In the

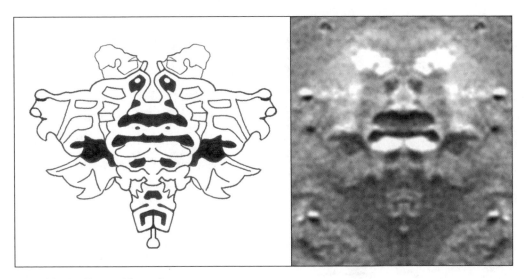

2.22 Martian Bat god (totem).

LEFT, Analytical drawing highlighting the winged Bat god and winged skull.

Drawing by George J. Haas

RIGHT, Mirrored half-faced geoglyph of Bat god on Mars. Note the bat-winged headdress.

From Dr. Mark Carlotto's Viking mosaic

Popol Vuh story the Bat god, who is associated with darkness and death, snatches off the head of the Hero Twins' father (First Lord).[12] Because of this association with decapitation the bat was called "he who tears off heads."[13]

2.23 Winged pectoral (Olmec, 1000 B.C.).
Note cleft in the forehead and wing shape of the flanged headdress.

Drawing by George J. Haas. (Image source: *The Blood of Kings*, by Schele and Miller, page 129.)

2.24 Stela F: Maya king Cauac Sky (dressed as the World Tree, Quirigua).

Drawing by George J. Haas. (Image source: *A Forest of Kings*, by Schele and Freidel, page 91, figure 2:15.)

On a 24-foot-high stela from Quirigua (Stela F (M6)), representing the Maya lord Cauac Sky, is a totem-like mask of the Bat god.[14] Included in this mask of the Bat god is a partial face of a bat skull that is masterfully incorporated into the elaborate headgear of Cauac Sky like a totem (Figure 2.24). The Maya king Cauac Sky was the same lord who sacrificed 18 Rabbit at Quirigua. He then positioned Smoke Monkey to accede to the throne of Copan.[15] As discussed earlier, it is the split faces of 18 Rabbit and Smoke Monkey that were found on the mesa just below the original Face.

A comparison of the two masks is provided in Figures 2.25 and 2.26.

The Jaguar Glyph

The next example of a profiled structure was found northeast of the Face, at the opposite end of the Viking mosaic from the Bat god. The original Face would be halfway between it and the D&M pyramid (Figure 2.12). This profiled landform is an enormous, segmented structure (more than 3 kilometers in length) that resembles a common Mayan pictograph of a jaguar-human (Figure 2.27).

The full image consists of three interlocking pieces whose top portion is the profiled head of a jaguar. An odd horseshoe-shaped feature forms an eye, while a highly reflective ridgeline forms

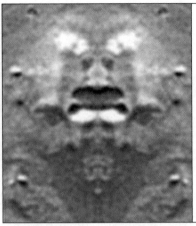

2.25 Martian Bat god. Note the short wings on the Bat god, and how his chin rests right on the skull below it.

2.26 Maya Bat god and bat skull (detail from Figure 2.24). Note the short wings on the Bat god, and how his chin rests right on the skull below it.

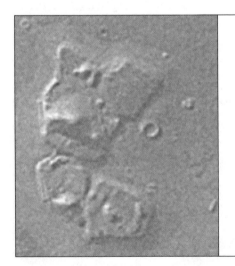

2.27 Half-faced jaguar structure (human-jaguar glyph).

From Dr. Mark Carlotto's Viking mosaic

2.28 Maya jaguar glyph. Note the paws or logographs below the head, the squat, square-shaped ear, and the three spots on the cheek.

Drawing by George J. Haas.. (Image source: *A Forest of Kings,* by Schele and Freidel, page 52, figure 1:3a.)

the profile of a muzzle and snout. Just below the eye are three small knobs or "spots" on the cheek area. These spots are similar to the three pelt marks that are often seen in the jaguar glyphs of the Maya. Behind the eye is a squat, block-shaped ear that sits on top of the face, just as it does in the glyph (Figure 2.28).

A large, profiled nose can be seen pressed against the edge of the mesa on the left side. The "snarling" presentation of the nose is similar to the snarling aspect in which the Olmec and Maya portray their Jaguar god. Between the head and the "paw" feature is a circular structure that is below the jaw line of the Martian jaguar. The circular structure and the paw seem to mimic the Mayan use of affixed logographs (glyph tags) in their hieroglyphic scripts. The Mayan glyph for jaguar has three logographs below a head of a jaguar that looks very much like the structure on Mars (Figure 2.28).

After a mirror flip along the left side of the profiled geoglyph is performed, the full image of a snarling jaguar-human face is observed (Figure 2.29A). Notice how the secondary structure, which mimics the Mayan logographs below the head, also comes to life when mirrored (Figure 2.29B). This secondary structure was unforeseen but, as you will find out in the following chapter, the discovery would become very significant.

These problematic landforms at Cydonia were starting to look more like a complex set of Martian geoglyphs just waiting to reveal their codex of secrets to the hearts and minds of the initiated. However, within all of these unexpected discoveries there was nothing more surprising than what was awaiting us right here on Earth, at a Maya site known as Palenque.

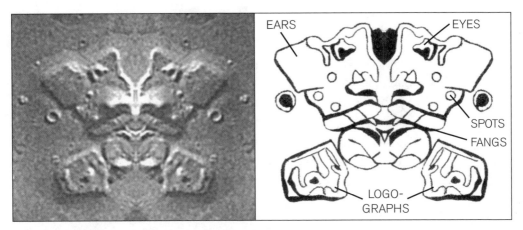

2.29A Martian jaguar glyph (half-faced jaguar structure, mirrored). Note the paws or logographs below the head, the squat, block-shaped ear, and the three spots on the cheek.

From Dr. Mark Carlotto's Viking mosaic

2.29B Analytical drawing of Martian jaguar glyph. Drawing by George J. Haas

Notes

1. Douglas Gillette, *The Shaman's Secret: The Lost Resurrection Teachings of the Ancient Maya* (New York: Bantam, 1997), 229. The Uay, which is also spelled Way or nawal, was believed by the Maya to be a companion-animal spirit into which sorcerers and kings were to transform themselves. See Schele and Mathews, *The Code of Kings,* 417.

2. J. Eric S. Thompson, *Maya Hieroglypic Writing* (Norman, Okla.: University of Oklahoma Press, 1971), 231.

3. Ibid., 230.

4. Linda Schele and David Freidel, *A Forest of Kings: The Untold Story of the Ancient Maya* (New York: Quill, 1990), 319.

5. Although these split-faced structures in the Viking mosaic were actually independently discovered earlier by George J. Haas between 1993 and 1994, the same bifurcated landforms were re-examined with the help of William R. Saunders in 1998.

6. Linda Schele and David Freidel, op. cit., 71.

7. Linda Schele and Peter Mathews, *The Code of Kings: The Language of Seven Sacred Maya Temples and Tombs* (New York: Touchstone, 1999), 73.

8. Maria Longhena, *Maya Script* (New York: Abbeville Press, 2000), 50. In Mayan inscriptions the turtle glyph represents birth. This idea is fostered by the creation story of the Maize god, who was reborn in the ball court. The Maya saw the ball court as a representation of the crack in the back of the Cosmic Turtle; the turtle therefore was connected to birth. See Schele and Mathews, *The Code of Kings,* 73.

9. Linda Schele and Peter Mathews, op. cit., 73.

10. E. Michael Wittington, ed., *The Sport of Life and Death: The Mesoamerican Ballgame* (Singapore: Thames & Hudson, 2002), 59. The word *hacha* is a Spanish term for axe.

11. On April 4, 2001 NASA released two MOC images of the "Third Face," M1400709 and M1500479. Unfortunately, both strips captured only a portion of the left side of the geoglyph, leaving the right side unresolved.

12. Irene Nicholson, *Mythology of the Americas* (London: Hamlyn, 1970), 185.

13. Jean Chevalier and Alain Gheerbrant, *A Dictionary of Symbols* (New York: Penguin Books, 1996), 71.

14. Linda Schele and David Freidel, op. cit., 316.

15. Ibid., 90.

CHAPTER 3

Hidden Faces

The Sarcophagus Lid of Pacal

In 1992, as NASA was preparing to launch the ill-fated *Mars Observer*, an independent scientist named Maurice M. Cotterell declared that he had broken the code of the famous Lid of Pacal (Figure 3.1). This fantastic sarcophagus was originally discovered in 1952 at Palenque by a Mexican archaeologist named Alberto Ruz. In recent years, the carved Lid of Pacal has become well known because of speculations about its connection with ancient space flight. Many believe the Lid depicts King Pacal working a control panel while sitting in what appears to be a flying vehicle. Skeptics say that this interpretation is pure fantasy.

Most scholars contend that the Lid depicts the king at "the moment of death as he falls from the world of living to the Underworld."[1] In Cotterell's unorthodox analysis of the inscriptions, he proclaimed that concealed within the complexity of the Lid's design is an inscribed matrix of the original *Popol Vuh* of the Maya. The *Popol Vuh*, considered to be the "Maya Bible," was thought to be written in Mayan hieroglyphics on accordion-folded books that documented the history of the Quiche Maya. Portions of the original text and those that had been transcribed into Spanish in the sixteenth century were saved from destruction when the Spanish burned many of the Maya records during their conquest.

3.1 Lid of Pacal. Note the "broken" corners of the Lid (upper right and left).

Drawing by George J. Haas. (Image source: Photograph by Merle Green Robertson.)

According to the first lines written in the *Popol Vuh:* "... The original book, written long ago, once existed but is now hidden from the searcher and from the thinker."[2] Cotterell interpreted this cryptic sentence to mean that perhaps the "hidden" images he was finding on the Lid of Pacal were actually the concealed pages from this hidden book. He discovered that through the use of transparent overlays on the Lid of Pacal, images of Maya gods could be found that paralleled the creation story of the *Popol Vuh.* He demonstrated that when the "broken corners" of the Lid of Pacal are over-lapped, with the aid of transparencies, the meaningless half figures and partial glyphs along its edge burst into life.[3]

When Cotterell performed an overlay of the top border boxes of the Lid of Pacal, a totemic set of faces emerged, including a bird and human head, followed by a snarling tiger and a dog head (Figure 3.2A). As you can see, the four faces were created out of three portraits of Maya lords (Figure 3.2B). According to Linda Schele, these three individuals have identities: a court official and two administrators, the latter two each identified as a "Keeper of the Holy Books."[4] In this context, where the Lid of Pacal is perceived as a multifarious matrix of a sacred book, the identity of these two lords as keepers of holy books extends tremendous support to Cotterell's belief that hidden images of the *Popol Vuh* are indeed incorporated within the design of this remarkable lid.

Cotterell found that the same results were achieved when the main body of the Lid was overlaid at different key points. It was not until these secret glyphs were matched up with their second halves (like a mirrored image) that it was possible to see what the glyphs were meant to represent. After a substantial number of the hidden images had been brought to light and identified by Cotterell, it became clear to him that the "breaking" of the corners was a deliberate and intentional part of the Lid's design.

One of the missing corners of the Lid had the remains of a broken glyph

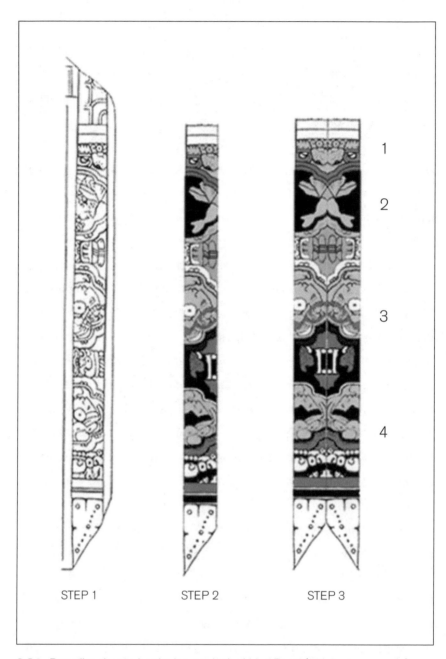

3.2A Decoding the top border images in the Lid of Pacal (with broken corners).

Overlay and colorizing by Jim Miller. (Image source: The Mayan Prophecies, by Cotterell, Appendix 8.)

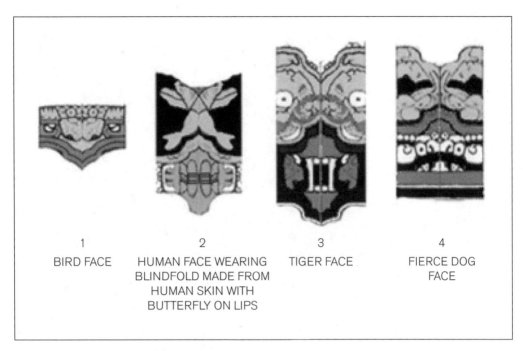

1	2	3	4
BIRD FACE	HUMAN FACE WEARING BLINDFOLD MADE FROM HUMAN SKIN WITH BUTTERFLY ON LIPS	TIGER FACE	FIERCE DOG FACE

3.2B Decoding the top border images in the Lid of Pacal (detail).

Overlay and colorizing by Jim Miller. (Image source: The Mayan Prophecies, by Cotterell, Appendix 8.)

that Cotterell called a defect marker. This marker was part of a crossed pattern of dots that would form a complete X if it were not broken. When Cotterell joined the corners of this marker through the use of acetate overlays, he not only restored the "defective" pattern, but the three Maya lords and the partial glyphs along the border boxes transformed into the group of stacked faces that are illustrated in Figure 3.2A. In the inner portion of the carving, he detected another defect mark on the ridge of the central character's nose. After placing the acetate drawing over the defect mark, he found that the image of the Maya Bat god miraculously appeared (Figure 3.3). Much to our surprise, Cotterell's overlaid image of the Maya Bat god looked a lot like the strange structure we had found on Mars that has a

3.3 Maya Bat god (Lid overlay). Note the alignment marker along the nose (under the chin of the bat).

Overlay and colorizing by Jim Miller (after Cotterell)

similar bat-winged headdress (Figure 3.4). Notice the wings that jut off the sides of the head in both images.

Although Cotterell's work with the Lid of Pacal has been widely criticized, we found his technique intriguing. With the help of anomaly hunter Jim Miller, we were not only able to duplicate the work of Cotterell, but we found many of his mirrored overlays to be located at significant alignments. Jim Miller duplicated the Maya Bat god in Figure 3.3 by rotating the Lid 20 degrees and then placing a transparent mirror image over the original drawing. The head of Pacal was butted together as a marker; this was established earlier by Cotterell. Miller found the rotation angle of 20 degrees to be very significant. To the Maya, 20 is the number of totality and signifies a complete K'atun of 20 years.[5]

So Cotterell was finding hidden faces carved in the stone sarcophagus of Pacal that were directly related to the myths of the Maya (and golden numbers)—and we were finding similar faces on Mars through the use of mirrored images! Compare the two images of the Bat god: Figure 3.3 found on the Lid of Pacal and Figure 3.4 found on the surface of Mars.

Among the many images that Cotterell found on the Lid of Pacal, the Jaguar god of Palenque was one of the most important because the jaguar

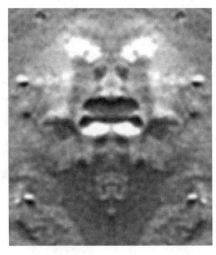

3.4 Bat god on Mars (mirrored image). NASA *Viking* image enhanced by Dr. Mark Carlotto

represents the fifth and present Sun of the Maya creation myth.[6] Initially, Cotterell identified all four of the previous ages of Creation on the Lid, but he could not find a signifier for the Fifth Sun. He found that the glyph signs for each of the ages were incorporated into the original design of the carving, and they did not require acetate overlays to be seen. However, the fifth and final age could only be found through an overlay of key points along the edge of the Lid.

When Cotterell overlaid the second dotted X pattern (which is actually a Sun sign) along the side border of another reversed transparency, the Jaguar god finally emerged. Like the many images that were revealed within the Lid of Pacal, the fifth age was also hidden from the casual onlooker. This single image of the Jaguar god would also become one of the most important faces to be found hidden in the Lid of Pacal. This image provides a major link between two totally independent investigations, ours on Mars and Cotterell's at Palenque.

3.5 Jaguar god (Lid overlay). Note the square ears, the eyes, the mouth, and the human-shaped nose.

Overlay and colorizing by Jim Miller (after Cotterell)

We found his overlaid image of the Jaguar god (Figure 3.5) to project an uncanny kinship to another one of our Martian discoveries, the human-jaguar glyph (Figure 3.6). Notice the overall contours of the Jaguar god's head in the overlay as compared to the mirrored jaguar structure we found on Mars. The common facial features are extraordinary—but the similarities do not end there!

The Cosmic Monster was another important symbol of the Maya, with a number of aspects to its symbolism. It personified the Milky Way as it stretches from east to west.[7] It also symbolized the Sun and Venus as they move through the heavens.[8] The Sun and Venus are synonymous with the

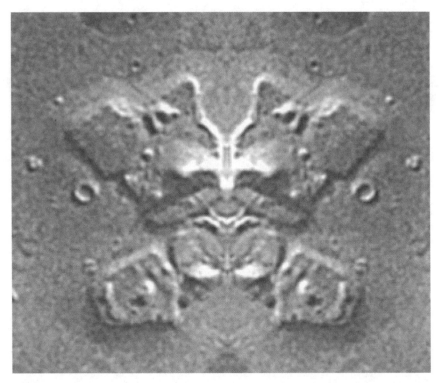

3.6 Jaguar glyph on Mars (mirrored image). Note the common features with the Lid overlay in Figure 3.5.

NASA *Viking* image enhanced by Dr. Mark Carlotto

twin gods First Lord and First Jaguar. The Cosmic Monster is depicted with a long, arched snout and clawed legs that are used to "thrash the sky as it surges around the universe."[9]

A temple mask depicting the front face of the Cosmic Monster (Figure 3.7) is found on the principal Royal Temple at Tikal.[10] This monstrous mask represents the planet Venus or First Lord.[11] The resemblance to that of Cotterell's overlay image beneath the Jaguar god is unmistakable (Figure 3.8). What Cotterell uncovered with his overlays was none other than the twin gods First Jaguar and First Lord stacked one on top of the other. As we have seen in Figure 3.5 and Figure 3.6, using a similar mirroring technique we

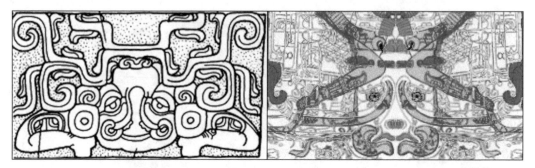

3.7 Front face of Cosmic Monster (Venus or First Lord) from royal temple at Tikal.

Drawing by Linda Schele, © David Schele. (Courtesy of Foundation for the Advancement of Mesoamerican Studies, Inc., www.famsi.org.)

3.8 Lower portion of Jaguar god overlay.

Overlay and colorizing by Jim Miller (after Cotterell)

uncovered the same image on Mars. Just as the original Face on Mars depicts the First Lord and First Jaguar, we now have a second structure that does the same!

The Mirroring Effect

At this point, a certain aspect of our investigation must be addressed: the anthropomorphic effect as an associative process of mirroring. When any surface is mirrored, even a random-patterned surface like wood grain, bilateral symmetry is instantly created. The brain wishes to make sense of the chaos and will focus on familiar aspects in the image that can be recognized. Strange mask-like faces can sometimes be found in these images that appear in a stacked formation, much like a totem. This phenomenon is called simulacrum, or false image. These types of masked and stacked faces are usually just one-sided, web-like contours without any substance or real depth.

False images can also be formed from folding over an inkblot on a piece of paper. This technique is used in psychoanalysis and is termed the

3.9 Aerial image of Earth.
(Image source: Monitoring Earth
Resources from Aircraft and Space-
craft, NASA, 1971, page 61.)

3.9A Left side mirrored.

3.9B Right side mirrored.

Rorschach Test, after its developer Hermann Rorschach. These types of images are considered projections and cannot form a complete proportional face. No matter how you look at them or fold them, these inkblots are just abstractions and will never conform to the right shape, size, and orientation of a real face. This process will never create a true split or half face as we have seen on Mars or in the Mesoamerican artifacts.

The aerial photograph in Figure 3.9 is an example of a typical rocky area found here on Earth. This photograph demonstrates some of the weird images and faces that are created when a random surface is mirrored. This aerial image of the Earth has been divided in half; mirror flips of each side are presented for your review (Figures 3.9A and 3.9B). Once the aerial image is mirrored through the center, an array of mask-like faces and other projections are created by the symmetry. The faces in this particular image are found to be disproportionate and tend to get lost in the web of geometric patterns. If the mirroring point is moved in any direction, all kinds of illusionary images may appear.

As mentioned, a similar mirroring effect with wood grain or other granulated surfaces can also produce these projections of geometric faces. The odd shapes and human-like faces that are produced are disproportionate or distorted in some manner and only subjectively correspond to something recognizable. How does one tell what formations are intended to be half images capable of being mirrored, such as the Maya artifacts shown thus far, and which formations are just false images and chance patterns? As you have seen and will see throughout this study, we consider the iconography of the original Face to be the archetypal model for which all other discoveries are judged. Therefore the criteria for accepted geoglyphs include a consistent design, style, and iconography that reflect a direct correlation between itself and adjoining geoglyphs.

Learning How to See

Donald M. Anderson, author of the book *Elements of Design* and a professor of art at the University of Wisconsin, states that for our eye to be able to recognize an object we have to understand the process of image formation. We perceive images as a whole instead of as isolated elements because of the principles of depth perception and the limits of image fusion.

For our eye to be able to fuse the pixels on a computer screen (or any picture) into a cohesive image, we need graduated values of black, white, and gray as well as the proper viewing distance. An image that may be seen quite clearly from 12 feet away will become indistinguishable if the image is enlarged several times and viewed too closely. The image's formation would become abstracted and we would only be able to perceive its separate pieces. This would result in the image being reduced to an indecipherable field of geometric patterns.[12]

The perception of the total image is formed by our eye's desire to organize things in an effort to form a complete image. In a black-and-white image, the eye searches for the visual characteristics, which are darker or lighter in contrast. The eye identifies similar shapes that are grouped closer to those in the rest of the viewing field, such as black shapes found within white areas giving form to the image. This type of image completion, through the separation of contrasting groups, is called "relative density."[13]

As an example of this visual phenomenon, Anderson uses a detailed image of an ancient ceramic mosaic pavement from the third century (Figure 3.10). As Anderson has suggested, if we focus on a small (black-and-white) section of the mosaic we will only see the meaningless textures and patterns formed by the geometric pieces of ceramic. At this resolution, the

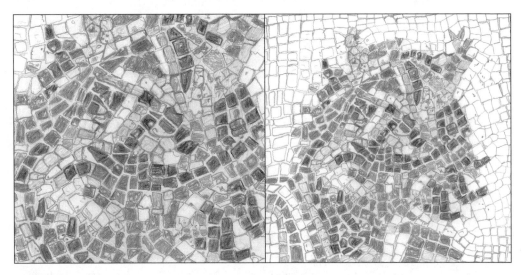

3.10 Mosaic pavement (detail, viewed at close range). Note the abstract textures and the odd patterns formed by the pieces of ceramic.

Drawing by George J. Haas. (Image source: Elements of Design, by Anderson, page 38.)

3.11 Mosaic pavement (detail showing Triton, Greek god of the sea).

Drawing by George J. Haas. (Image source: Elements of Design, by Anderson, page 38.)

bits and pieces of the ceramic act as cells or halftone dots used to create an image in newspapers or the pixels of digital images. When the full image of the object is registered in its proper viewing distance, we have no trouble identifying the bust of the Greek god of the sea, Triton (Figure 3.11).

To produce the proper fusion of any given image, it is important to designate the distance at which the image or structure was intended to be viewed. As with the problematic landforms at Cydonia, this distance remains ambiguous and varies depending upon which structure and which portion of the structure is being viewed. Scale is also a major problem when trying to form a cohesive image. Considering that the Face on Mars is an intentionally designed work of art that is a mile wide and a mile and a half long, we have to determine a proper viewing distance. When the 1976 *Viking* orbiter snapped the first image of the Face, it was 1,000 miles above the surface of Mars and as a result the camera captured a provocative portrayal of a human face. As the viewing distance is shortened with the 1998 MOC image (approximately 275 miles), this incredibly designed structure reveals a different visage—a split-faced mask.

In Dr. Tom Van Flandern's preliminary analysis of the origins of the Face on Mars, he put forth the hypothesis that the proposed architects who built it may have intended the structure to be viewed from a nearby planet which has since exploded.[14] Dr. Van Flandern contends that Mars was once a moon of a now missing planet and the structure known as the Face was meant to be viewed from this home planet at a precise distance. He believes that its facial configurations could have been designed on such a scale that it would only be recognized as being a face at the proper optimal viewing distance.[15]

Once the distance is established for any given image and the object comes into view, it's our cultural influences that take over. These cultural influences can include such unique iconographic motifs as design, style, and graphic symbolism. As a viewer, how are we to understand the structural configurations of any object when we may not know what we are seeing? Like the hidden faces within the Lid of Pacal, there are sometimes ambiguous elements that can be concealed within an unfamiliar image that appear to be hidden from the casual onlooker.

A classic example of a double image, called Rabbit/Duck, is presented as an optical illusion of two different portraits within one face (Figure 3.12). Depending on your point of view and personal perception of the drawing, the face may look like a rabbit (facing to the left) or a duck (facing to the right). This perceptual effect is known as "contour rivalry."[16] As readily apparent in this example, the technique of contour rivalry allows one image to intentionally have two different readings.

A well-established artist who experimented with the Idea of hidden and double images was Pablo Picasso. Considered the most extraordinary artist of the twentieth century, Picasso developed with his close

3.12 Rabbit/Duck. Do you see a rabbit (facing left), or do you see a duck (facing right)?
Drawing by George J. Haas

friend Georges Braque a system of art called cubism. Inspired by the sculptures of African tribesmen, they began this almost scientific process of painting (circa 1909) by breaking down the individual aspects of a subject into their most essential elements. This cubist technique allowed facial features to be seen through a combination of multiple viewpoints, with an endless array of angles and fractured planes presented all at once. This fragmentation of the subject established a new vision of space and movement within a two-dimensional plane. No longer was the viewer restricted by seeing a face from only a single vantage point.

In a later and simpler style of cubism developed by Picasso, a typical

3.13 Picasso painting: Seated Woman (detail). Note that the left side of the face is in profile and the right side is a frontal view.

Drawing by George J. Haas. (Image source: William Rubin, ed., Pablo Picasso: A Retrospective, New York: The Museum of Modern Art, page 413.)

face could be seen in both a frontal view and in profile simultaneously within a single portrait. This split-faced technique was such a common feature in much of Picasso's later work that it became a signature style, as we see in his Seated Woman of 1953 (Figure 3.13). This bifurcated process was explored further throughout Picasso's career.

Figure 3.14 exemplifies the key aspects of this cubist technique that complement the many two-faced attributes we have found on Mars. This synthesis of split faces is found in a painting titled Italian Girl, which is derived through another, more geometric, version of cubism. In this detail we readily see a classic example of a split-faced composition that combines the duality of the full face with the profile.

The left side of the portrait is a human profile of a very masculine face with a dark, almond-shaped eye. The right side represents a more feminine, organic form by depicting a frontal view of a round, pumpkin-like head, complete with a square eye and a dangling earring. The basic two-faced composition of Picasso's Italian Girl has a lot in common with the pictorial language found in Mayan iconography, most specifically within the design of glyphs (Figure 3.15).

Notice how the strong male profile on the left side of the glyph over-

3.14 Picasso painting: Italian Girl
(detail). Note how the male profile
on the left overlays the pumpkin-like
female head on the right.

Drawing by George J. Haas. (Image
source: Picasso, by Diehl, page 35.)

3.15 Maya glyph: lunar series
(reversed). Note that the male profile
on the left overlays the round, formed
pumpkin head on the right.

Drawing by George J. Haas. (Image
source: Maya Hieroglyphic Writing,
Thompson, Figure 37, No. 22.)

lays the soft, round, pumpkin-like head seen on the right. The similarities
between Picasso's cubist style of portraiture and the two-faced design found
in Mayan glyphs are amazing. It should be clear from these few examples
that the cubist technique of presenting multiple viewpoints within a single face has kept the idea of two-faced portraiture alive and well in modern art. Almost single-handedly, Picasso established the art of the two-faced
process as a legitimate and historical technique.

Another example of how we interpret what we see is demonstrated in a
painted face (Figure 3.16) that was produced by the Huichol Indians in
Mexico. In this example we see how our perception of an image is also
determined by our understanding of the context in which the image is presented. This information aids in our recognition of the complete image.
Before reading on, please take a good look at this painting.

3.16 Huichol face painting (Mexico). Note the hidden images within the decorative marking on the face.

Drawing by George J. Haas. (Image source: Elements of Design, by Anderson, page 36.)

When examining the painting, we are sure you will have no trouble seeing the fine linear contours that form the shape of the face. However, the blobs of paint that appear splattered around the face will seem to form meaningless shapes. These blobs demonstrate the main idea of image formation because their recognition depends on knowing what to find within the cultural context of any image. Now, if you look again at the face painting and are instructed in what to observe, you may see something completely different. If you look closely you will notice that the blobs of splattered paint that appear to be just decorative markings are actually stylized renderings of animals. On the left side of the face, from the cheek area to the chin, six standing deer can be found. A full dog can be seen on the right side of the chin and there are six winged birds along the forehead.[17] As you can see by the presence of the hidden animals in the Indian painting, learning how to see and being instructed in what to see is an essential element in understanding what you are looking at. As with the hidden faces of the sarcophagus Lid of Pacal found at Palenque, we have to know what to examine before we can see it.

It must be emphasized at this point that the optical illusions illustrated in the last few examples are international works of art. They are not "tricks of light" or examples of simulacrum. They are to be regarded as premeditated designs whose meaning is, however, sometimes subjective to cultural bias.

With regard to our capacity to identify blobs as shapes of animals within the painted face of the Huichol Indians, we realize that the eye also has the tendency to find faces in clouds and piles of laundry. It also has the propensity to see profiles of famous faces on mountain boulders and deformed potato chips. However, these types of faces are only seen in contoured representations of profiles and have no substance. These naturally formed images can generally be seen from only a single vantage point and will disappear as the viewer physically moves around it. None of these types of effects are compliant with any of the images we have found on Mars or in Mesoamerica.

Mirrors in Mesoamerica

From what we have learned so far at Palenque, it appears that the Maya had an immense understanding of the elements of image fusion; this understanding allowed them to produce the hidden faces found within the elaborate design of the Lid of Pacal. It stands to reason that they also must have understood the anthropomorphic effects of the mirroring process. The fact is that the Maya (and many other Mesoamerican cultures, including the Olmec) produced finely crafted mirrors that are of exceptional quality. The mirrors of the Aztec are primarily made out of polished obsidian and pyrite. However, one example of a marcasite mirror has a surface ground with such precision that it produces a magnified reflection.[18]

Earlier cultures such as the Olmec were also skilled in producing highly polished concave mirrors of the finest quality that included such metallic materials as magnetite, hematite, and pyrite.[19] Scholars believe that the technology used to produce these mirrors is extraordinary considering the so-called "primitive" culture in which they were used. For decades scientists have been amazed that the curved surfaces of these lenses are ground to such perfection that even after inspecting them with a microscope, no evidence of abrasives can be found.[20] In a manuscript titled "Concave Mirrors from the Site of La Venta, Tabasco," Robert F. Heizer and Jonas E. Gullberg of the University of California state:

> The degree of polish of the concave surfaces [of these mirrors] is remarkably good, and seems to represent the limit of perfection which the material will allow. No clear trace of abrasion marks on the polished surface can be detected.... The effect is almost identical to the modern practice of parabolizing optical reflectors.[21]

Although scholars cannot agree on the purpose served by these ancient optical instruments, they do agree that these are some of the finest mirrors ever made.

3.17 Montezuma with the magical mirror in Codex Florentino (mid-sixteenth century). Drawing by George J. Haas. (Image source: Aztecs of Mexico, by Vaillant, Figure 4, page 243.)

As recorded in the Aztec book known as the Codex Florentino, just prior to the Spanish conquest a strange bird (which had a mirror in its head) was found by Aztec hunters and brought to the court of Montezuma (Figure 3.17). As Montezuma peered into the magical mirror he first saw the vastness of the heavens and then saw visions of Hernando Cortes and his invading army.[22]

Were these actually prophetic mirrors linked to some type of ancient "remote viewing" used by the Maya? According to Mayanist Linda Schele, these sacred mirrors were not only used for prophecy; they were also used to open portals where ancestors and gods would materialize.[23] If these magical mirrors could actually open portals and foretell the future, maybe they were also used as the matrix for an exercise in image projection. Archaeologist Richard A. Diehl suggests that these finely crafted optical lenses were used as "camera lucida" instruments, allowing scribes to project images at will.[24]

The Maya technique of mirroring the half profiles and split images that are concealed within the design of artifacts is not unique to the Lid of Pacal. Another example of this method of design can be seen on a small Maya sun disk (Figure 3.18), which actually has a mirror in its center. This Maya sun disk, one of only three such disks that have been discovered, was found inside a throne seat in the Temple of Chac Mool at Chichen Itza.

In the center of this turquoise mosaic disk is a golden pyrite mirror; it is surrounded by eight mosaic compartments.

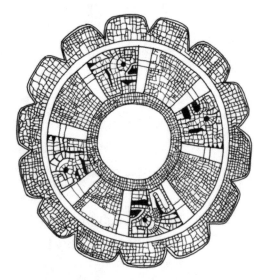

3.18 Mosaic sun disk (Temple of Chac Mool). Note the four half-faced heads of the Feathered Serpent around the disk.

Drawing by Linda Schele, © David Schele. (Courtesy of Foundation for the Advancement of Mesoamerican Studies, Inc., www.famsi.org.)

Profiles of the Feathered Serpent or Quetzalcoatl (Kukulcan) alternate around the central mirror.[25] If one places a mirror along the edge of the image, the profile becomes a frontal presentation of the Feathered Serpent (Figure

3.19). This same idea of reflecting half images of faces that we saw on the border of the Lid of Pacal is also incorporated in the border of this mirrored Sun disk, this time portraying the Feathered Serpent.

In Aztec mythology, Quetzalcoatl's adversary was called Tezcatlipoca, which means "Smoking

3.19 Mirror flip of the Feathered Serpent.

Mirror." *Tezcatli* means mirror and *poca* means smoke.[26] According to Aztec myths, Tezcatlipoca was seen as the personification of evil and prophecy.

It is said that Tezcatlipoca, while in disguise as a turkey, flew too close to the horizon and one of his feet was amputated by Earth Monster. Tezcatlipoca had his foot replaced by a mirror that could see the innermost thoughts of men.[27] In an Aztec drawing of Tezcatlipoca (Figure 3.20), notice the disk shape of the mirror that replaces the ankle and the "smoking" forms flowing out of the mirrors that occupy the headdress and foot area. These elements form the glyph for Smoking Mirror.

With this many references to mirrors, it is not surprising that the Maya also had a glyph for mirror. The mirror glyph (Figure 3.21) is represented as a half image: a D-shaped mirror.[28] The grooved design around the mirror makes it look as if it has a cut or beveled edge. To the authors, the flat edge along the one side of the glyph suggests a first-hand knowledge of the mirroring technique. Because of the shape of this mirror glyph, it invites the action of mirroring. We don't believe this is wishful thinking. As it turns out, we are not the first to ponder this idea.

3.20 Aztec god Tezcatlipoca (Smoking Mirror). Note the smoking mirror on the left foot and the mirror on the pectoral medallion.

Drawing by George J. Haas. (Image source: Daily Life of the Aztecs, by Carrasco and Sessions, page 51.)

3.21 Mayan glyph for mirror. Note that the mirror glyph is presented as one half of a circular disk.

Drawing by George J. Haas. (Image source: Blood of Kings, by Schele and Miller, Plate 111c, page 283.)

Herencia (Cut in Half)

The first archaeologist to suggest that Mesoamerican artifacts had been designed with the option of mirroring was Jacinto Quirarte. His discovery occurred during an investigation comparing Izapan-styled iconography with that of Olmec and Maya art. In Quirarte's paper "Tricephalic Units in Olmec, Izapan-Style and Maya Art," he contends that if the profiled head of an Izapan Avian Serpent glyph (Figure 3.22) were "duplicated" (or mirrored) at the tip of the snout, the half profile would then appear on both sides, forming a single head.[29]

3.22 Avian Serpent glyph (from a Tlapacoya ceramic vessel).

Drawing by George J. Haas. (Image source: After Joralemon, 1976: Figure 6c, Olmec & Their Neighbors, by Coe and Grove, org., page 295.)

This experiment resulted in a complete frontal view of an Avian Serpent being produced, similar to the one seen on an incised jadeite object found at Rio Pasquero (Figure 3.23).

A second archaeologist to explore this mirroring idea, Anatole Pohorilenko, presents a great example of a "cut-in-half" artifact in her paper "The Olmec Style and Costa Rican Archaeology."[30] She presents an example of the half face of a young Olmec man cut into a large jade pendant (Figure 3.24). This odd artifact is classified as a "herencia," which simply means that an object is cut in half.[31] The half mask is perfectly split through the forehead, including the nose and mouth. Again, the only way to complete the image is by mirroring. However, because of the recognizable subject matter, no mirroring is needed here. It should also be noted that because of the precise placement of the suspension hole just above the eye, the pendant hangs perfectly straight.[32]

This remarkable Olmec pendant is comparable in design to a half-faced mask produced hundreds of years later by the Kwakiutl Indians of North

3.23A Avian Serpent glyph, mirrored. Note the completed mask.

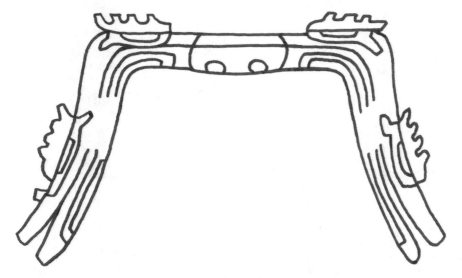

3.23B Avian Serpent incised on a jadeite object (Middle Formative period). Note the common features of the horizontal head, at the top of the bar, with the mirrored Avian Serpent glyph in 3.23A.

Drawing by George J. Haas. (Image source: After Joralemon, 1976: Figure 8, Olmec World: Ritual and Rulership, by Coe, ed., page 84.)

America (Figure 3.25). Just as we see in the Olmec mask, the Kwakiutl mask is half-faced and carved in a full frontal presentation that is split in half. The presence of this North American Indian mask further emphasizes the common iconographic motifs shared by the indigenous people of the Americas.

3.24 Olmec half-faced mask (jade pendant).
Note that the face is not in profile; this is a frontal view that has been cut in half.

Drawing by George J. Haas. (Image source: Olmec & Their Neighbors, by Coe and Grove, org., page 319.)

3.25 Kwakiutl half-faced mask (British Columbia, circa 1880).

Drawing by George J. Haas. (Image source: Cultures of Native North Americans, by Feest, page 447.)

Terrestrial Geoglyphs

The unusual nature of the design of the Cydonia landscape has caused other researchers to consider its structures to be highly eroded. We think that the clarity of the images presented in this book indicates that this is not the case. Although there seems to be some deposition of sand or dust, the effect of erosion appears to be minimal.

Although one needs a mirror to decode many of the hidden geoglyphs at Cydonia, they still exist as half or split images on their own. Even with their vastness of scale, they hold their own when compared to a common terrestrial iconography presented in the artwork of the Maya and the Olmec.

As you will see, the possibility that this "Martian Codex" was created by some random set of geologic processes is beyond the laws of probability. We must also add to the equation the fact that these bifurcated Martian geoglyphs, have not only been found to be directly connected with a common iconography shared with the cultures of Mesoamerica, but each half-face is intimately related to its counterpart via the same myths.

We find the relationship to be overwhelming. It establishes a testable data set (on Mars and Earth) that supports our contention that the geoglyphic images on these structures at Cydonia are indeed artificial. The question in most minds will be, if these structures are artificial, what is the reason behind their construction? Surely they are more than mere "calling cards" for interplanetary travelers.

From the Martian surface they may not appear to be the type of aesthetically pleasing monuments one would expect to find in a modern city or metropolis, such as the Eiffel Tower, the Statue of Liberty, or even the Taj Mahal. Their immense size and the fact that, in most cases, they can only be viewed from high above a normal human's visual plane dictate that they must have been constructed for some other optical purpose.

Speaking to something far beyond the geometric grid layout of streets and structures found in our modern cities, Martian architecture may be more in tune with such enigmatic earthworks as those found among the ancient cultures of both North and South America. Throughout North America the indigenous people who were once referred to as the Mound Builders built a variety of earthworks in the shape of animals and geometric forms that number in the thousands.[33] Recent surveys of the mysterious platforms and mounds built by the Olmec at San Lorenzo, Mexico revealed that when combined the protruding masses reveal an enormous stylized jaguar mask staring at the skies above.[34] Likewise in South America travelers have strolled across the Nazca plains for centuries, unconscious of the sacred text they were defiling. A vast network of intermingling straight lines, circles, spirals, and giant figures of animals went unseen by these "uninitiated" travelers for decades. It wasn't until the 1940s, when they could be seen from the air, that the hidden images of the Nazca plateau were finally realized.

Dr. Paul Kosok, a professor at Long Island University, was the first to study the line drawings in Peru. Kosok declared that the entire set of these pictographic drawings was like a lost book. He referred to the drawings as "the largest astronomy book in the world."[35]

Moving farther inland, one finds more gigantic geoglyphs that were built to be viewed from high above the ground. At the time of the Spanish conquest, the Inca capital of Cuzco was found to be in the shape of a giant puma. To the Inca the puma was seen as a symbol of strength and power, and they constructed the city of Cuzco in the form of a crouching puma.[36] Its shape is still visible today within the modern city of Cuzco (Figure 3.26). Virtually the entire topography surrounding Cuzco is littered with structures that mimic the form of animals. At Ollantaytambo there is a temple on the flanks of the Tamboqasa mountain that represents a giant llama, while a portion of the massive complex at Machu Picchu represents a lizard.[37]

Farther to the north, beyond the city of Lima, are the ruins of Caral, located in the Supe Valley. From recent excavations of this site, some archaeologists are hailing this almost forgotten complex as the home of the earliest known settlement in the New World; they date it to well before 2600 B.C.[38] Just beyond this ancient complex of mounds and half-buried pyramids is an immense half-faced stone geoglyph (Figure 3.27) set into the surface of this once-sacred ground.

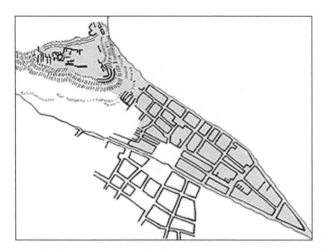

3.26 The puma-shaped city at Cuzco. Note the highlighted shape of the puma.

Drawing by George J. Haas. (Image source: Monuments of the Incas, by Hemming and Ranney, page 43.)

3.27 Half-faced geoglyph (Caral, Peru, 2500 B.C.). Note the small eye, the bulbous nose, the gaping mouth, and the raked hair.

Drawing by George J. Haas. (Image source: Smithsonian, August 2002, Vol. 33, no. 5, page 64.)

Notice the D-shaped head with its large gaping mouth and raked hair. It should be noted that this partial face is not carved in profile—it is designed in a "cut-in-half" manner, just as we saw with the Olmec half-faced pendant in Figure 3.24. The only difference is that, like the two-faced geoglyphs found on Mars, the Caral face is meant to be seen from high above the ground.

Our voyage between Mars and Earth concludes with the visionary work of one of the twentieth century's most important Japanese American sculptors, Isamu Noguchi. The Noguchi Museum, which is located just beyond the Queensboro Bridge in New York,[39] brings us full circle by leading us right back to Mars. Within the museum's vast collection of the artist's life work is a small model of an unrealized monument entitled "Sculpture to be seen from Mars."

In 1947, with the horrors of the atomic bomb fresh in his mind, Noguchi envisioned the construction of a gargantuan monument that would survive the annihilation of the human race. His proposed sculpture was conceived as an abstract geometric rendering of a human face over a mile long (Figure 3.28). Unlike the immense geoglyphic monuments seen on Mars, his face was never constructed.

3.28 Sculpture to be Seen from Mars. Model, 1947, by Isamu Noguchi.

Photograph by Soichi Sunami. Reproduced with the permission of the Noguchi Museum, New York

From what we have discovered on Mars, perhaps Noguchi wasn't the only sculptor-architect in our solar system to have contemplated the idea of constructing a visual "marker" establishing a link between these two neighboring planets. Are the pictographic mounds and structures found throughout the New World the remnants of an ancient Martian analog right here on earth? Could it be that the geoglyphic structures found at Cydonia were once bustling metropolises, not in the shape of animals, but in the shape of half and bifurcated faces?

The very existence of these earthworks establishes that a conscious effort was made by both ancient and modern builders to communicate to a higher plane. The figures project toward the sky, just as we see with the gigantic geoglyphs at Cydonia on Mars.

In addition, when one is confronted with the Olmec half-imaged artifacts, the hidden faces within the Lid of Pacal, the Mayan mirror glyph, and the prophetic god Smoking Mirror, it becomes overwhelming evidence of the importance of mirroring to these New World cultures. Did the makers of these hidden and half-carved images want us to perform mirror flips of these works of art? Did they want us to complete the message and read the pages of a sacred book that was "hidden from the searcher and thinker?" If so, what would they tell us? The answer to these assertions may lie in the very design of Mayan art and a composite writing system that is based on a complex matrix of faces.

Notes

1. Frank W. Porter III, *The Maya* (New York: Chelsea House, 1991), 59.

2. Adrian G. Gilbert and Maurice M. Cotterell, *The Mayan Prophecies: Unlocking the Secrets of a Lost Civilization* (Rockport, MA: Element, 1995), 76.

3. Ibid., 68–84.

4. Linda Schele and Peter Mathews, *The Code of Kings: The Language of Seven Sacred Maya Temples and Tombs* (New York: Touchstone, 1999), 111–112. Schele professes ignorance as to why such lords would be represented on the sarcophagus Lid of Pacal; she only offers speculation that they may have been included as architects and overseers of the construction of the temple and tomb. Considering Cotterell's research suggests that the Lid is a holy book, Schele may have been more right than she thought.

5. To recreate the overlays of the Lid of Pacal by Maurice M. Cotterell, Jim Miller used Paint Shop Pro, version 7 (PSP7). Jim Miller provides the following explanation of his process:

   For all the images I utilized a drawing of the sarcophagus Lid of Pacal rendered by George Haas. All images were overlaid and colorized using a general opacity of 50 to 85 percent. Rotating the Lid of Pacal 20 degrees (the Mayan number of totality) and then placing a transparent mirror image over the original created the Bat god. The head of Pacal was butted together at the nose as a marker. Again I colorized one side, did a crop and a second mirroring of the half image to achieve the end result.

   The Jaguar god came about by placing the Lid of Pacal on the horizontal (of the left-hand border) and placing a mirror image over the original and matching up the "X" (sun glyph) at the center, just as instructed by Cotterell. The colorizing process remained the same as [for] the Bat god.

I have compared my work to Cotterell and I have found that the basic drawings are quite different. I would not have achieved the same results had I used all of his markers of alignment. His base drawing appears to be a bit wider and a trifle longer overall. I also compared George Haas's drawing to Cotterell's and an additional one I found on a German website as well as the original Lid of Pacal itself and found that the Cotterell drawing is the one with the most differences. My conclusion is that the Haas drawing is a more accurate rendition of the lid. This, however, doesn't reduce Cotterell's work in its value and the fact that I was able to reproduce the various "hidden images" does indicate that even with his slight distortions, the images are actually there.

I selected 20 degrees for my rotation of the Bat god as an end result of toying with the images. I was unaware of previous rotational degrees used when I began this project and did do a bit of research to discover what was important mathematically to the Maya. Richard Hoagland uses 19.5 as a basis for a great deal of his work regarding links between Mars, NASA, Masons, etc. So this number sprang to mind as I began. In my readings I came across the totality number of 20 and when I applied it to the Lid of Pacal, it worked best in duplicating Cotterell's findings for the Bat god. I found that many of Cotterell's images could be duplicated by rotating the Lid at common degree points, such as 18, 20,33, 52, 64, and 72.

6. Adrian G. Gilbert and Maurice M. Cotterell, op. cit., 320.

7. Linda Schele and Peter Mathews, op. cit., 410.

8. Linda Schele and David Freidel, *A Forest of Kings: The Untold Story of the Ancient Maya* (New York: Quill, 1990), 115.

9. Douglas Gillette, *The Shaman's Secret: The Lost Resurrection Teachings of the Ancient Maya* (New York: Bantam, 1997), 32.

10. Linda Schele and David Freidel, op. cit., 169.

11. Douglas Gillette,op. cit.

12. Donald M. Anderson, *Elements of Design* (New York: Holt, Rinehart & Winston, 1961), 38–39.

13. Ibid., 36.

14. Tom Van Flandern, *Dark Matter, Missing Planets & New Comets: Paradoxes Resolved, Origins Illuminated* (Berkeley: North Atlantic Books, 1993), 438.

15. Tom Van Flandern, "Preliminary Analysis of April 5 Cydonia Image from the *Mars Global Surveyor* Spacecraft," *Meta Research,* 1999. www.metaresearch.org

16. Rebecca Stone-Miller, *Art of the Andes: From Chavin to Inca* (New York: Thames & Hudson, 1995), 42.

17. Donald M. Anderson, op. cit., 36.

18. George C. Vaillant, *Aztecs of Mexico: Origin, Rise, and Fall of the Aztec Nation* (Garden City: Doubleday Doran and Co., 1941), 142.

19. Michael D. Coe, *The Olmec & Their Neighbors: Essays in Memory of Matthew W. Stirling* (Washington, D.C.: Dumbarton Oaks Research Library and Collections, 1981), 110.

20. Editors of the Reader's Digest, *Mysteries of the Ancient Americans* (Pleasantville, NY: The Reader's Digest Association, 1986), 135.

21. Michael D. Coe, op. cit., 111.

22. George C. Vaillant, op. cit., 242–243.

23. David Freidel, Linda Schele, and Joy Parker, *Maya Cosmos: Three Thousand Years on the Shaman's Path* (New York: Quill, 1993), 175.

24. Richard A. Diehl, *The Olmec America's First Civilization* (London: Thames & Hudson, 2004), 94. The camera lucida uses a prism and magnifying lens attached to a frame to project an image onto paper.

25. Linda Schele and David Freidel, op. cit., 394.

26. Richard Cavendish, *Mythology: An Illustrated Encyclopedia* (New York: Barnes & Noble, 1993), 252.

27. Irene Nicholson, *Mythology of the Americas* (London: Hamlyn,1970), 226.

28. David Freidel, Linda Schele, and Joy Parker, op. cit., 140–141.

29. Michael D. Coe, op. cit., 296.

30. Anatole Pohorilenko offers a second example of a "cut in half" image incised on an Olmec bar pendant. See Michael D. Coe, op. cit., 320.

31. Ibid.

32. Ibid., 318.

33. George E. Stuart, "Who Were the Mound Builders?" *National Geographic* 142, no. 6, (December 1972), 790.

34. Henri Stierlin, *Art of the Maya From the Olmec to the Toltec-Maya* (New York: Rizzoli, 1981), 31.

35. Editors of Reader's Digest, *The World's Last Mysteries* (Pleasantville, NY: Reader's Digest Association, 1978), 282.

36. *Incas: Lords of Gold and Glory* (Alexandria, VA: Time-Life, 1992), 51.

37. Fernando Elorrieta Salazar and Edgar Elorrieta Salazar, *Cusco and the Sacred Valley of the Incas* (Lima, Peru: Ausonia S.A., 2003). A photograph and illustration of the llama-shaped Temple is on pages 104–105. A photograph and illustration of the of the Lizard complex is on pages 134–135. In addition, the authors document the existence of many more structures that are built in the shape of animals throughout the area.

38. Ruth Shady Solis, Jonathan Haas, and Winifred Creamer, "Dating Caral, a Pre-Ceramic Site in the Supe Valley on the Central Coast of Peru," *Science* 292, no. 5517 (April 27, 2001), 723–726.

39. The Noguchi Museum is located at 33rd Road in Long Island City, New York.

CHAPTER 4

Two-Faced

Duality

Why build an array of profiles or half structures and scatter them across a vast area of Mars? The answer may lie in the basic simplicity of a binary system. It would seem that the original Face on Mars is the keystone to the whole complex of Cydonia. Because of its asymmetrical facial features, the Face introduces us to a binary system of split faces and a theory of transmutation or transformation.

Biologically, transmutation is the change of one species into another. In geometry, it is the change of one figure into another of the same area, such as a circle into a square.

The Sphinx-like appearance of the Face holds the template of the opposing forces of nature, which is at the root of our existence. This ancient mystery of duality is manifest in the Face and is repeated in the surrounding structures in the Cydonia complex. This idea of duality is as old as humankind and forms the lexicon of our common symbols. The archetypal pattern of the nature of duality encompasses all opposites such as man/woman, day/night, fire/water, round/square, and life/death.

This idea can be compressed into the basic makeup of a single seed that splits into two halves. The halves become the twins of the self, thereby giving birth to the third component. Simply put, when opposites are united, their union produces a new hybrid offspring of both species. According to J. E. Cirlot, duality can be described as follows:

> Duality is a basic quality of all natural processes in so far as they comprise two opposite phases or aspects ... [which] can be either symmetrical or asymmetrical.[1]

The idea of duality is traditionally related to the ancient symbol of the Gemini, which is expressed in the twin gods Castor and Pollux. Symbolically, the Gemini are often represented as two pillars. They can be seen as the Pillars of Hercules or as the columns of Solomon's Temple, called Jachin and Boaz. The two architectural pillars, like the Gemini, are symbols of the twin. Most important, these pillars are symbols of the primordial and androgynous Gemini.[2]

Despite being the twin sons of Zeus, Castor and Pollux were complete opposites. The first twin, Castor, was mortal and of the Earth. The second twin, Pollux, was immortal and a god of the heavens.[3] To understand the opposition and inversion principles of this myth, we look again to Cirlot. He explains that the first pillar is identified as Castor, the Earth Twin who personifies the split of opposites such as the two-headed Janus. The second pillar is recognized as Pollux, the Heavenly Twin who expresses opposites. This twin of opposites is then fused together, creating Oneness.[4]

The myth of Gemini is the essence of opposites, but what we have not mentioned is that it is also a symbol of inversion. According to Marius Schneider, the idea of inversion holds a morphology that we are formal components of varying and unvarying aspects: one half of our being speaks to our individuality and the other half links us with our species. He says the continuity of life is assured by mutual sacrifice, which is consummated on the twin peaks of the mystic mountain, where death permits birth.

In essence, all opposites are fused together for an instant and then inverted. What is constructive becomes destructive: love turns to hate, evil to good, unhappiness to happiness, and martyrdom to ecstasy.[5] The megalithic conception of this mystic mountain, according to Schneider, is described as follows:

> The mountain of Mars (or Janus) which rises up as a mandorla of the Gemini is the locale of Inversion—the mountain of death and resurrection; the mandorla is a sign of Inversion and interlining, for it is formed by the intersection of the circle of the Earth with the circle of heaven. This mountain has two peaks, and every symbol or sign alluding to this "situation of Inversion" is marked by duality or by twin heads.[6]

This idea of twin heads is embraced within the image of Janus, referred to as the gatekeeper of time (Figure 4.1). His two faces look in opposite directions, toward the past and the future, denoting both awareness of history and foreknowledge, this being an allusion to the dualism of time. The Roman god Janus was also the god of doorways and passages and was placed at the threshold of temples.

The origins of Janus are in the Greek god Orthus, who was the two-headed watchdog of Geryon. Orthus was the father of the Sphinx, who held the riddle of the ages, which was concerned with time and the transformation of man. Orthus was also seen as the equivalent of Sirius, the Dog star.[7]

Another example of back-to-back facial duality that may not be as familiar as the Roman god Janus is found in ancient Mesopotamia. The Sumerian god Isimud (Figure 4.2), thought to be

4.1 Janus (Roman coin). Note that Janus was the god of gates and doors.

Drawing by George J. Haas. (Image source: *Larousse Encyclopedia of Mythology*, by Aldrington and Ames, page 200.)

the forerunner to both Orthus and Janus, also had two faces. Isimud, known as the two-faced gatekeeper of heaven, was also personal minister of the Sumerian water god Enki.

Similar to what we have seen in these earlier examples of two-faced deities, the Olmec of Mesoamerica also had a concept of duality. On a small jade plaque dated to 1150 B.C., we find two identical Olmec profiles looking in opposite directions (Figure 4.3). This comes close to matching the presentation of the Roman god Janus. Notice the flaring upper lip, which is a typical snarling aspect of the Olmec were-jaguar. A star-shaped glyph is carved above the central void area where the two snarling heads meet, giving the whole object a celestial connection. The glyph has been identified as Venus and the void as a Cosmic Portal.[8]

Like the Romans, Sumerians, and Olmec, the Maya also had a god with two faces that looked in opposite directions. Known as Kukulcan to the Maya and Quetzalcoatl (the Feathered Serpent) to the Aztecs, this god was sometimes presented as a two-faced god. The name

4.2 Isimud (Sumerian two-faced god).

Drawing by George J. Haas. (Image source: *Mythology: An Illustrated Encyclopedia,* by Cavendish, page 87.)

4.3 Janus face (Olmec jade plaque, 1150 B.C.).

Drawing by George J. Haas. (Image source: *The Olmec World: Ritual and Rulership,* by Coe, ed., page 258.)

Quetzalcoatl has the dual meaning of bird (*quetzal*), and serpent (*coatl*).[9] The word *coatl* can also mean twin, thus Quetzalcoatl is also known as the Magnificent Twin.

Quetzalcoatl was regarded as the single embodiment of the Maya twin gods First Lord (Venus) and First Jaguar (Sun). An elaborately carved Huaxtec sculpture shows a standing Quetzalcoatl looking in two directions while wearing a conical ocelot skin cap. The frontal face of the god (Figure 4.4) is seen as the bright star of Venus, while the back portrays the face of the

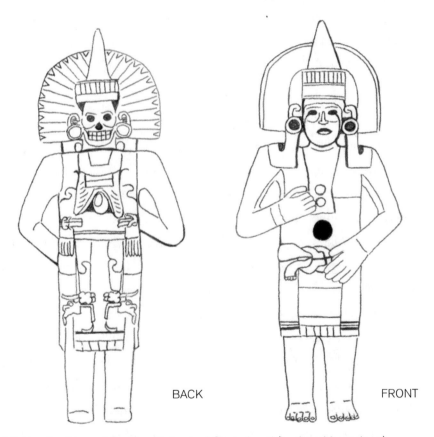

BACK   FRONT

4.4 Huaxtec figure of the Maya/Aztec god Quetzalcoatl (back and front views). Drawings by George J. Haas. (Image source: *Mexican and Central American Mythology,* by Nicholson, page 81.)

Sun.[10] In Mesoamerican cultures the idea of paired gods and oppositions was at the heart of cosmological thought. These oppositions function as metaphors for the transformational change of one thing into another.[11]

In the Aztec myth of creation, based on a much earlier Maya-Olmec story, a supreme celestial being called Ometeotl (the Lord of Duality) is the source of all creation. According to this myth Ometeotl (Hunabku in Maya tradition) is the supreme god of all dualities:

> He/she was the lord of all beings, the one who embodied and dom-
> inated the contrary forces of the cosmos. He/she encompassed all
> dualities, including those of masculine and feminine, life and death,
> and spirit and matter. His/her abode was in the place of duality, the
> thirteenth heaven that rose above all the other heavens.[12]

The Severed Head

To anyone familiar with the pictorial characteristics of Mayan hieroglyphs, it soon becomes apparent that a significant number of them are based on the use of half and profiled faces. Unlike the Egyptians, whose pictorial signs contain almost no isolated faces, the Maya developed a pictographic system that incorporates an almost endless array of eloquently character- ized faces. These expressive pictographic faces are fashioned in the genre of a severed man's head. The isolated image of a severed head was seen as a central symbol of royal power and sacrifice among the kings of the Maya.[13]

In the Maya creation myth, it is the original Hero Twins who are decap- itated by the Lords of Death after being tricked into losing a ball game. One of the twins is buried in the center of the ball court and the other has his head hung in a gourd tree as a warning to all others.[14]

This decapitation idea goes back to the beginnings of Mesoamerican culture, which begins with the Olmec. The Olmec produced colossal sculp- tures of individual severed heads that have been found buried throughout

Veracruz, Mexico (Figure 4.5). The most common trait shared by these exotic heads, aside from the Negroid facial features,[15] is that they wear a headdress that looks like protective headgear. This tight-fitting headdress resembles the ball-court helmets worn by the Maya. In these ball games, scholars believe that the losers were beheaded in the center of the ball court and the heads were displayed as trophies.[16]

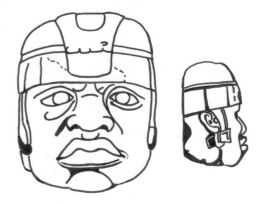

4.5 Olmec colossal head. Note the crossed eyes and flatness of the rear of the head.
Drawing by George J. Haas. (Image source: After Juarez, Sanchez, and Rey, *Olmecs: Arqueologia Mexicana*, inside fold-out cover.)

This interpretation of the Olmec heads as being displayed as severed heads is reinforced by the fact that most of the eyes appear to be crossed (Figure 4.5), which is a signifier of decapitation. Another feature of these sculptures is that many of the heads are flat and featureless on the posterior side, perhaps suggesting that they were meant to be resting on the backs of their heads, looking up to the sky.

Maya Glyphs

The ritual of decapitation played a major role in Mesoamerican religion; it also provided a rich lexicon of faces that were assembled into a complex pictographic language. The variety of half and profiled faces used by the Maya includes animals, birds, serpents and other grotesque and mythological creatures. The images in Figure 4.6 are typical examples of a few of the profiled heads used in Mayan glyph design. Besides single heads, the Maya would overlap two profiles, creating a double-headed glyph that often combined both human and animal faces (Figure 4.7). This complex system of writing is truly unusual.

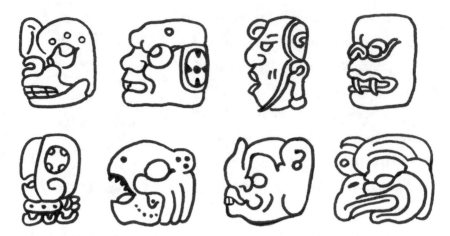

4.6 Maya head glyphs (profiled heads).

Drawing by George J. Haas. (Image source: *Maya Hieroglyphic Writing*, by Thompson, various pages.)

4.7A Double-headed glyph. Human with grotesque head.

Drawing by George J. Haas. (Image source: *An Introduction to the Study of the Maya Hieroglyphics*, by Morley, page 183, figure 69.)

4.7B Double-headed glyph. Human with bird head.

Drawing by George J. Haas. (Image source: After Linda Schele image number 4067.)

The foundation of the writing system is believed to have been developed by the Olmec, transformed by the Maya, and then absorbed into the Aztec lexicon. Evidence of its influence can be traced throughout the Americas. Although these profiled glyphs are part of an unfolding Mayan hieroglyphic language and their exact meaning is still debatable among Mayanists, the archetype for many of these profiled faces has been recognized as head variants. An assortment of these alternative profiles can be used to replace graphic signs for numbers and directional signs within inscriptions. Other glyphs that feature faces and a vast array of graphic forms are used as phonetic signs, which act like logographs.

The main problem in understanding the Maya system is that many of their glyphs have variant meanings that include changes of phonic and semantic values. Further complicating their meaning, many of these glyphs can morph into different graphic forms and styles such as "sign compounds" and "converging glyphs."[17] The merging of two distinct signs creates new meanings, of which only a small portion are fully understood. These compound glyphs are the result of a process called conflation.[18] Linda Schele expounds on this problem:

> The complexity of the system is often bewildering to the modern reader, just as it must have been to the ancient Maya who was not an expert in its use. But we must recognize that the goal of the writing system was not mass communication, in the modern sense.... Writing was a sacred proposition that had the capacity to capture the order of the cosmos....[19]

After examining the facial quality of these glyphs, we believe their meaning goes far beyond these current debates and may be connected to the Face on Mars. Similar to the split-faced Sphinx design of the Face (and the surrounding landforms), we have discovered that the Maya not only incorporated the use of half and profiled heads in their inscriptions; they also utilized the technique of the split face in their sculpture.

Split-Faced Sculpture

The following set of artifacts reveals just a portion of a vast cache of split-faced sculptures that were produced throughout the cultures of Mesoamerica. The first example is a fragmented head of a Zapotec child that was associated with Pitao Bezelao, the god of death. This split face (Figure 4.8) is classified as a funerary head chronicling the eternal process of human destiny from the vitality of youth to the disintegration of death in one startling image.[20] When the image is split and each side mirrored, the left side features a sullen-faced child wearing a frown (Figure 4.8A). The right side has a skull-like face displaying a distinct grin (Figure 4.8B).

Another example of the Mesoamerican perception of duality is found in a striking pottery mask from a grave at Tlatilco, Mexico (Figure 4.9). One half of this bifurcated mask is a human face with a protruding tongue (Figure 4.9A). The companion side features a skull that appears to be feline, possibly a jaguar (Figure 4.9B). The feline skull has a prominent set of teeth and displays a knob feature on the side of the head that looks like the remnants of an ear. Although whom these separate faces represent is debatable, the split-faced design is unmistakable.

4.8 Zapotec funerary head. Drawing by George J. Haas. (Image source: *National Museum of Anthropology, Mexico City*, by Ragghianti and Collobi, 1970, page 85.)

4.8A Left side mirrored (Life).

4.8B Right side mirrored (Death).

In the spring of 1996 an extensive exhibit of Olmec sculptures was held at the National Gallery of Art in Washington, D.C. A special issue of *Arqueologia Mexicana* available at the exhibition was devoted exclusively to the art of the Olmec. It contained one of the most amazing examples of Mesoamerican sculpture demonstrating the use of a split image found anywhere. It is an odd sculpture of an Olmec head that resides at the Tenochtitlan Community Museum in Veracruz.

We discovered a split face here that had been completely overlooked by other researchers (Figure 4.10). According to scholars at the museum in Veracruz, this sculpture is an intentionally mutilated head. They believe the left side of this strange sculpture is the remnants of an intact Olmec head, while the right side has been obliterated.

The article contends that the facial features on the right were intentionally eliminated by drill holes and deep perforations, leaving a deformed

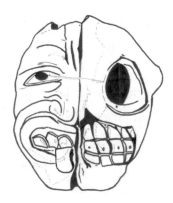  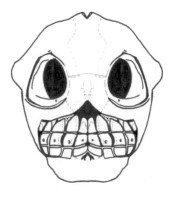

4.9 Human and jaguar skull mask.

Drawing by George J. Haas. (Image source: *The Flayed God: The Mythology of Mesoamerica*, by Markman and Markman , page 52.)

4.9A Left side mirrored (human face with protruding tongue).

4.9B Right side mirrored (feline skull).

4.10 Olmec mutilated head (stone sculpture).
Drawing by George J. Haas. (Image source: *Olmecs: Arqueologia Mexicana*, by Cyphers, page 58.)

and abstract morphology. Some scholars believe that this Olmec head was originally a dualistic representation of two faces, but for some unknown reason one half was removed.[21]

Something didn't feel right about this explanation. In examining the overall shape of the sculpture, it became apparent to us that the "mutilated" side was not proportional to the side with the head. If this head had been a symmetrical face at one time, why was there a greater mass of stone on the left side? There had to be another answer.

We performed a mirror flip of each side of this anomalous Olmec sculpture. The right side resembled the typical colossal Olmec head that can be found throughout the plains of the Gulf of Mexico, only on a smaller scale (Figure 4.10B). The features of the face are very Negroid, with thick lips and a broad nose. We also noticed that this face had no eyes. Instead of eyes it had what appeared to be eye shields or eyeglasses. If you look closely, you can see a line cut into the side of the face right where the frames would have been for the eyeglasses. In addition, on the side of the head (out of view) was a four-fingered hand with long fingernails covering the ear.

The whole face, with the hands covering the ears, the pushed-back open mouth, and the eye shields gave the impression of someone squinting while looking at the bright light of an explosion!

After performing a mirror flip of the left side of this object, a stunning image of a complete open-winged bat was staring back at us (Figure 4.10A).

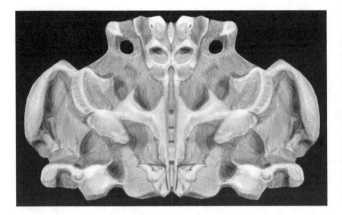

4.10A Left side mirrored (spread-winged bat). Note the fox bat shape of the face and the butterfly shape of the bat's tail.

4.10B Right side mirrored (human face). Note the cleft in the forehead and the rectangular shape of the eye socket.

The side of the Olmec sculpture that was supposed to be a mutilated mess was actually the half image of a bat!

Imagine looking down onto the Cydonia landscape and seeing a mesa with this type of bifurcated topography. Because of its intangible construction and asymmetrical morphology, it would be difficult to determine if this visage was intentionally sculptured or just an oddly eroded mesa.

The bat image on the so-called mutilated side of the head is not the simulacrum effect of mirroring. The split image of an Olmec head and an open-winged bat is an intentionally sculpted work of art. The sculpture is a key example of the Mesoamerican codex of duality. It may actually represent the Olmec Bat god, much like the winged head seen in Chapter 2. How many other camouflaged artifacts like this Olmec head have been misinterpreted and left in the basements of museums?

In our view, this Olmec sculpture of the split human-bat image, like the other Mesoamerican artifacts presented here, isn't any different than the half and split images we have found on Mars. In fact, their source of inspiration may be directly linked!

Notes

1. J. E. Cirlot, *A Dictionary of Symbols* (New York: Barnes & Noble, 1995), 24.

2. Ibid., 116.

3. *Brockhampton Reference Dictionary of Classical Mythology* (London: Brockhampton Press, 1995), 52.

4. J. E. Cirlot, op. cit., 116.

5. Ibid., 24.

6. Ibid., 117.

7. Robert Graves, *The Greek Myths* (Wakefield, RI: Moyer Bell, 1994), 130.

8. Michael D. Coe, *The Olmec World: Ritual & Rulership* (Princeton: Princeton/Abrams, 1996), 268.

9. Irene Nicholson, *Mexican and Central American Mythology* (New York: Peter Bedrick, 1985), 81.

10. Richard Aldrington and Delano Ames, *The Larousse Encyclopedia of Mythology* (New York: Barnes & Noble, 1994), 431.

11. Linda Schele and David Freidel, *A Forest of Kings: The Untold Story of the Ancient Maya* (New York: Quill, 1990), 416–417.

12. Roberta H. Markman and Peter T. Markman, *The Flayed God: The Mesoamerican Mythological Tradition: Sacred Text and Images from Pre-Columbian Mexico and Central America* (San Francisco: HarperSanFrancisco, 1992), 53.

13. Linda Schele and David Freidel, op. cit., 124.

14. Ibid., 74.

15. Editors of Reader's Digest, *Mysteries of the Ancient Americans: The New World Before Columbus* (Pleasantville, NY: Reader's Digest Association, 1986), 132.

16. Editors of Reader's Digest, *The World's Last Mysteries* (Pleasantville, NY: Reader's Digest Association, 1978), 264.

17. Personal conversation with Marc Zender at the Twentieth Annual Maya Weekend, Maya K'atun Celebration, University of Pennsylvania Museum, April 5–7, 2002. Handout: Marc Uwe Zender, "Problems Encountered in Decipherment of Logographic, Syllabic and Logosyllabic Scripts," 2002, 2.

18. Ibid.

19. Linda Schele and David Freidel, op. cit., 55.

20. Carlo Ludovico Ragghianti and Licia Ragghianti Collobi, *Great Museums of the World: National Museum of Anthropology, Mexico City* (New York: Newsweek Inc. and Arnoldo Mondadori Editore, 1970), 85. The Maya and Aztec produced a variety of bifurcated sculptures and masks depicting similar life and death motifs.

21. Ann Cyphers, "The Colossal Heads," *Olmecs* (*Arqueologia Mexicana*, 1996), 58–59.

CHAPTER 5

An Inverted Face on the Horizon

The City Was Gone

A mere two weeks after the release of the first MOC strip of Cydonia in which the Face was captured, the world was about to be offered two additional swaths of unusual Martian real estate to digest. NASA had publicly announced that it would attempt to target the area known as the City Square and the City Center Pyramid with the second swath of Cydonia scheduled for April 14. This is an area to the southwest of the Face that was shown to have numerous large structures resembling pyramids in the 1976 *Viking* images. Richard Hoagland had suggested that this area was once a city, complete with a City Center or City Square. Hopes ran high for confirmation that the pyramids were artificial. Unfortunately, the camera missed the City Center completely, only to photograph an area about a kilometer or two to the west (Figure 2.2). Thus, while the controversy surrounding the "cat box" incident was still brewing, concerned researchers were presented with yet another setback.

Although this rustic portion of the "city" had many unusual features and odd geometric-shaped mounds, it held little interest to anomaly researchers, let alone the public. The consensus among concerned followers was a feeling of disappointment, and not long after the third Cydonia swath was taken and released, the second swath was quickly set aside. The true purpose of this second so-called "miscalculated image" would, however, eventually be revealed to us. In our opinion, it proved to be the result of an intentional target by NASA. It would also become a cornerstone of our whole investigation. (More about this in Chapters 9 and 10.)

On April 23, 1998, when the Cydonia region was photographed for the third and final time in this series, the orbital camera hit pay dirt. This time the City Center was captured right on target (Figure 2.3). Although the intended objective was finally achieved, something about this new image of the city complex just didn't look right. The main pyramid and surrounding objects looked more like deflated mounds than actual pyramids (Figure 5.1). Where was the city?

If this collection of odd formations, which NASA termed buttes and hills, were indeed structural remnants of a city, then their construction was surely not pyramidal. What were thought to have been organized pyramids within the city now appeared to be heavily mantled mounds covered with Martian sand. The structures looked nothing like the richly formed pyramids spotted in the *Viking* images by

5.1 Context image–bottom portion of third Cydonia swath (SP1–25803).

Taken by the Mars Orbital Camera (MOC) on board the *Mars Global Surveyor* spacecraft (*MGS*). Courtesy of NASA/JPL/Caltech

Richard Hoagland; rather, they were an assortment of shapes and sizes with odd conjunctive features. It appeared that the city was gone.

As a result of NASA's view that all three swaths taken during the targeting of Cydonia showed nothing but natural features, the mainstream media concluded the story was truly over. However, for us the Martian Codex was just beginning to reveal its secrets.

A Bat in the Cross Hairs

The feature that Richard Hoagland had termed the City Square was a cluster of four mounds that, on the 1976 *Viking* image, appeared to form a square. He pointed out that the four large oblong mounds, forming a square or diamond shape, were like "cross hairs" where one could look east and see the Sun rising over the Face on Mars.[1]

The new, higher-resolution images from the MOC revealed that the largest of these four mounds had extended zigzag ridges and facets that seemed to flow out from its sides. This portion of the cross hairs that Hoagland had seen was actually a structure approximately 400 meters in length, looking very much like the half image of an open-winged bat (Figure 5.2).[2] When the structure was mirrored down the central ridge, a complete bat could be seen with its wings spread wide open (Figure 5.3). If one compares this mirrored bat image on Mars with the mirrored Olmec bat sculpture from the previous chapter, some amazing similarities emerge (Figure 5.4).

As we have seen, the bat was a major symbol in Maya culture and was considered the embodiment of death and darkness. One of the most intriguing similarities between these two images is the "butterfly" at the bottom center of both bats. While the bat is a symbol of death and darkness, the butterfly is symbolic of life and rebirth. It is also an ancient emblem of the soul, signifying an "unconscious attraction toward the light."[3] The duality

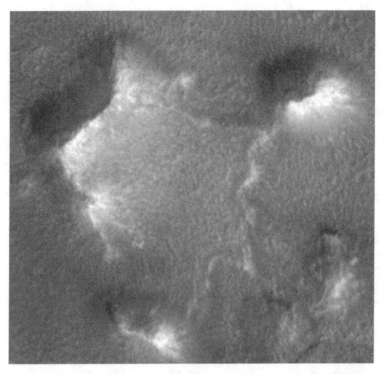

5.2 Bat in the cross hairs (portion of third Cydonia swath known as the City Square). Note the half image (geoglyph) of an open-winged bat.

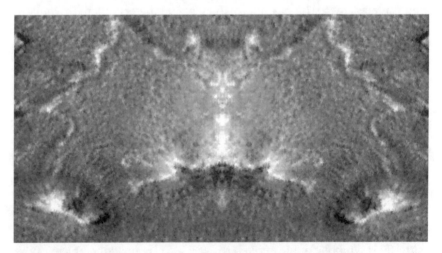

5.3 City Square bat (mirrored). Note that additional confirmation of the City Square bat is available in MOC image E0500156.

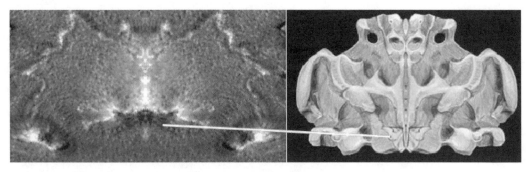

5.4 Bat comparison (mirrored Olmec bat and mirrored City Square bat on Mars). Note the butterfly image at the bottom center of both images.

of death and resurrection are symbolized in this mirrored image of bat and butterfly!

We believe the bat is the dominant image because it plays a dual role here. A bat, whose nature is to hang upside down, is also a symbol of inversion.[4] With this in mind, we found it very interesting that a symbol of inversion would be found at the very center of the city complex. The strategic placement of the bat in the cross hairs was significant because it provides an alignment with the Face. As Hoagland had pointed out, it is from this location that one could have stood at a right angle to the Face and looked out across the landscape to see a solstice sunrise appearing out of the mouth of the Martian Sphinx.

Hoagland not only determined that a solstice sunrise could be seen from here, but calculated that approximately 450,000 years ago the heliacal rising of Earth would also occur, moments before the solstice sunrise.

The Face Inverted

In the first chapter we showed how the lower panels on the Maya temple at Cerros identified the mask as the Jaguar Sun on the horizon. We also referred to Hoagland's connection of the Egyptian god Horus to the Face.

He found that the term "Horus on the horizon" could also mean "Face on the horizon." Just as the Jaguar Sun was the god of the rising and setting Sun to the ancient Maya, Horus was the god of the rising and setting Sun in ancient Egypt. The setting Sun symbolized death, while the rising Sun symbolized resurrection.

The ancient texts of Egypt refer to the "Hawk god," Horus, who was the son of the goddess Isis and the god Osiris. These were three of the most worshipped gods throughout Egyptian history. Isis was seen as the embodiment of the star Sirius, while Osiris was associated with the constellation Orion. In his book *The Orion Mystery: Unlocking the Secrets of the Pyramids,* Robert Bauval describes how he discovered that the positions of the three main pyramids of Giza, Egypt (Cheops, Chephren, and Mycerinos) correspond to the three stars in the belt of the constellation Orion.[5]

In 1964, Alexander Badawy discovered that the two shafts in the so-called king's chamber of the Great Pyramid were aligned with Alpha Draconis, which was the pole star in 2600 B.C., and the three belt stars of Orion, "to help the soul of the dead king to rise to the special starry heaven of Sahu-Osiris (Orion)."[6] In 1986, Bauval determined that in the same period (2650–2600 B.C.), the queen's chamber would have pointed to the star Sirius (Isis), the Dog star, in the constellation Canis Major.[7] Celestially, the three stars in the belt of Orion point directly to the star Sirius. The heliacal rising of Sirius was extremely important to the ancient Egyptians, and many temples were built to honor this moment. The temple of Isis at Denderah, Egypt was so precisely aligned that at the exact moment of the rising of the star, light from only the star itself would beam through the main corridor to illuminate the main altar. A hieroglyphic inscription from the temple states:

> She shines into her temple on New Year's Day, And she mingles her light with that of her father Ra on the horizon.[8]

The ancient Egyptian holy city of Edfu is dedicated to the god Horus. He is said to have established a foundry of "divine iron" where he kept "the great winged disk that could roam the skies." Egyptian text declares: "When the doors of the foundry open, the Disk riseth up."[9]

Using the bat's inversion symbolism as a clue, we are now about to see an amazing example of contour rivalry, which was introduced in Chapter 3. To view the split image of the Face on Mars in Figures 1.9 and 1.10, we used

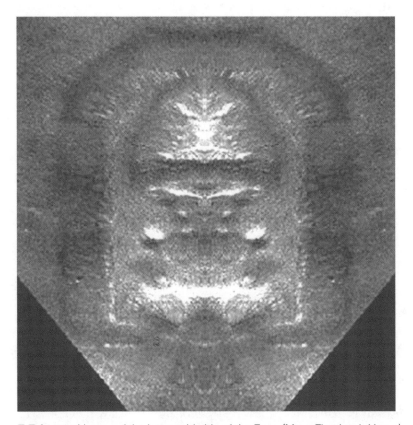

5.5 Inverted image of the humanoid side of the Face (Maya First Lord–Venus). Note the images of Sirius (Dog star), Horus (Falcon), Ra (Sun), and Isis (Mother goddess). The Sun, in the center, depicts Ra, the Sun God, defines Horus as the god of the Rising Sun, and acts as the solar-disk headdress of Isis. (See highlighted image, Figure 5.6.).

a contrast-inverted image from the original NASA photo. This contrast inversion allowed us to see the depiction of the Maya First Lord as Venus, and First Jaguar as Jaguar Sun. However, if we now use the original contrast and just invert the images—that is, turn them upside down—what appears is a depiction of the heliacal rising of the star Sirius, the Dog star (Figure 5.5)!

The first intriguing aspect of this inverted image to warrant our attention is an array of odd pictographic features that act as facial ornaments on the upright version—the mask of the Maya First Lord (the humanoid side). Within these objects a whole set of recognizable pictographic elements begins to emerge. The pictograph within the chin ornament on the mask of the Maya First Lord becomes the star Sirius in the form of a tattered-eared dog. The semicircle of the star rising above the "Vault of the Sky" can be seen from the dog's cheek to the center of the outstretched arms. Directly below is a dark arc that is seen as the Egyptian Vault of the Sky, which domes the path of the Sun. The feature, which is the upper lip of the Maya god now transforms into a feathered emblem of Horus "resurrecting" the Sun from the Underworld (Figure 5.6).

In Egyptian mythology, Horus-Behdety, who was one aspect of the god Horus, transforms into a winged disk called Great God, Lord of Heaven.[10] The images in Figure 5.7 show a comparison with an Egyptian emblem of Horus-Behdety, the winged globe.

The inverted nose ornament takes on the celestial shape of the Sun god Ra, and acts as the solar headdress of Isis, who is found directly below with her hands raised in glorious praise to the heliacal rising of Sirius. (Note: If you look closely at Figure 5.5 you can see the thumb and fingers of Isis.) Her long skirt is shaped by the inverted W (the tri-leafed crown emblem) of the Maya First Lord, whose Teardrop feature becomes her hand.

Upon closer inspection, Isis appears to have an ankh symbol hanging from each wrist. A second set of hands also appears to emerge from behind

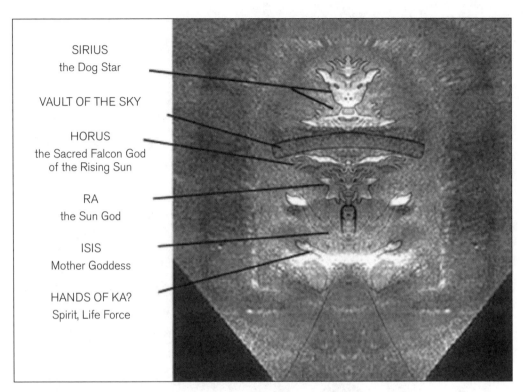

SIRIUS
the Dog Star

VAULT OF THE SKY

HORUS
the Sacred Falcon God
of the Rising Sun

RA
the Sun God

ISIS
Mother Goddess

HANDS OF KA?
Spirit, Life Force

5.6 Highlighted pictographics from Figure 5.5 (heliacal rising of Sirius). Note Sirius (Dog star), Horus (Falcon), Ra (Sun), Isis (Mother goddess), and possibly the hands of Ka (Spirit or Life Force) emanating from behind the skirt of Isis. If you look closely, it appears that Isis has an ankh hanging from her wrists.

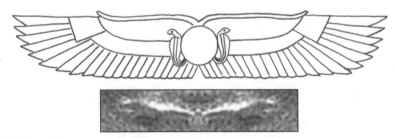

5.7 Winged Horus.

TOP, Nibiru (Marduk) (and symbol of Horus-Behdety).

Drawing by George J. Haas. (Image source: *History of Civilization: Earlier Ages,* by Robinson, page 56.)

BOTTOM, Mirrored image of the Horus pictograph (on the humanoid side of the Face on Mars).

the dress of Isis. The extended open palms, which conform to the shape of the eye sockets of the Maya First Lord, may be the mystical symbol Ka, meaning spirit, double, or life force. In the Egyptian creation myth, after Ra spews out the first twin gods, Shu and Tefnut, he places his arms around them so that their Ka will stay within them.[11] This mirrored image on Mars is very similar to a number of images found on tomb walls and in the Egyptian Book of the Dead. The image in Figure 5.8 shows the goddess Isis and the Ka with arms raised to the Sun God, Ra.

Another important Egyptian deity was Anubis (Figure 5.9), son of Osiris and Nephthys, adopted by Isis. He was depicted as dog-faced or jackal-faced and was known as the god of the dead. Anubis has also been associated with the Dog star, Sirius, and it was his job to determine if a dead soul could enter the court of Osiris. Anubis had a counterpart in the form of another jackal-headed deity known as Wepwawet or Upuaut, "the opener

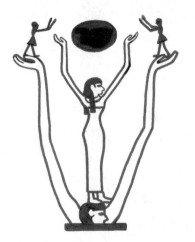

5.8 Isis and Nephthys (with the arms of Ka).

Drawing by George J. Haas. (Image source: *Great Ages of Man: Ancient Egypt,* by Casson and editors of Time-Life Books, page 81.)

5.9 Anubis (Egyptian god associated with the Dog star Sirius). Note the large fanned ears.

Drawing by George J. Haas. (Image source: *Tutankhamen,* by Desroches-Noblecourt (Boston: Penguin Books, 1989), page 221, figure 131.)

of ways." As Bauval says, they were "probably different aspects of the same divine archetype."[12]

Each of the inverted pictographs discovered hidden within the facial ornaments of the humanoid side of the Mars Face also has a Mesoamerican counterpart. The character of Anubis is synonymous with the Mesoamerican dog-headed god called Xolotl. Xolotl is known as the Divine Dog and, as with Anubis, has a counterpart, a twin or double: Quetzalcoatl (the Feathered Serpent), who represents Venus as the morning star.[13]

Xolotl is often presented as a death spirit in the Aztec Codex: the one who guides the dead on their journey to the Underworld. During certain times of the year, Xolotl carries the Sun to the Underworld;[14] it dies below the horizon as Venus rises in the guise of Tlahuizcalpantecuhtli, the evening star.[15] Xolotl also accompanies Quetzalcoatl in his descent to the Underworld to retrieve the bones of mankind. He then assembles the bones of the dead and restores them to life, much in the same manner that Anubis resurrects the dismembered body of Osiris.[16]

In the *Popol Vuh*, a dog plays a major role in the Hero Twins' first act of resurrection. While in the court of the Lords of Death, the Hero Twins are asked to sacrifice a dog. After they kill the dog, the Hero Twins immediately resurrect him. The dog is so grateful at being brought back to life that he begins to wag his tail.[17] Notice the jagged, tattered shape of the ears on the Martian pictograph of the canine depiction of the star Sirius. In many of the Aztec Codex books, the god Xolotl is commonly displayed with tattered or ragged ears.[18]

Just as the Egyptian god Anubis is associated with the twin star recognized as Sirius, it appears that the famous Mesoamerican dancing dogs in Figure 5.10 may depict more

5.10 Dancing dogs (ceramic, Jalisco, 200 B.C.). The positioning of the dancing dogs may mimic the rotational orbit of the twin stars known as Sirius.

Drawing by George J. Haas. (Image source: Postcard from The Art Museum, Princeton University, New Jersey.)

than just a pair of carefree canines. Xolotl is no different than other Mesoamerican gods that have complex attributes and opposing aspects. First of all, the very word *xolotl* means twin or twinned object. In the creation of the Fifth Sun, the Gods of Death chase after Xolotl in an effort to kill him. To escape their assault, he transforms into double objects such as a double maize plant, a double maguey plant, and a double salamander.[19]

So it should be no surprise to find the god Xolotl in an alternative aspect as double dancing dogs, representing the Dog star Sirius. It is our contention that the Mesoamerican sculptures of the so-called dancing dogs mimic the intimate rotational orbit of the twin stars known as Sirius A and B (Figure 5.11), in an emblematic depiction of two dogs dancing around each other. With what we have learned so far about the significance of the

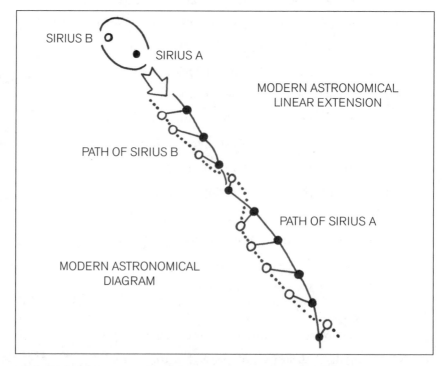

5.11 Orbit of Sirius stars A and B. Note the "dance" between the two stars. Drawing by George J. Haas. (Image source: *Sirius Mystery*, by Temple, page 69.)

Dog star Sirius and the parallels between the Egyptian god Anubis and the Mesoamerican god Xolotl, our interpretation of the dancing dogs seems quite plausible.

The Maya also depicted Xolotl as a single dancing dog with a tattered ear (Figure 5.12).[20] The single dog is seen here engaged in a triumphal dance as he emerges from the Underworld. Compare the jagged, tattered shape of the dog's ears on the Maya dancing dog and the Martian pictograph in Figure 5.5.

A dancing dog emerging from the Underworld would be an apt description of the Dog star Sirius rising in the morning—the same occurrence that was so highly celebrated by the ancient Egyptians. Robert Temple writes:

5.12 Maya dancing dog (detail of vase MSII20). Note the tattered ear, the position of the paws, and the raised foot signaling the dance.

Drawing by George Haas. (Image source: *Painting the Maya Universe: Royal Ceramics of the Classic Period*, by Reents-Budet et al., page 209.)

> The first appearance of Sirius on the horizon just before the sun— after seventy days in the Duat (Underworld)—was what is called the heliacal rising of Sirius. This event occurred once a year and gave rise to the Sothic Calendar.... (So this was the ancient Egyptian name for Sirius as it was spelled by the Greeks).[21]

Beneath the dog pictograph in Figure 5.6 is a horizontal band that connects the Earth and sky. The Vault of the Sky is to the ancient Egyptians what the skyband is to the ancient Maya (Figure 5.13). In Mayan iconography, the skyband is seen as a framing element that contains glyphs for the Sun, the Moon, and the planets, which are divided into individual segments within the band.[22]

Just below the skyband is the winged globe that can be compared to the Maya emblem of the Celestial Bird, or the Itzam-Ye. The Itzam-Ye is some-

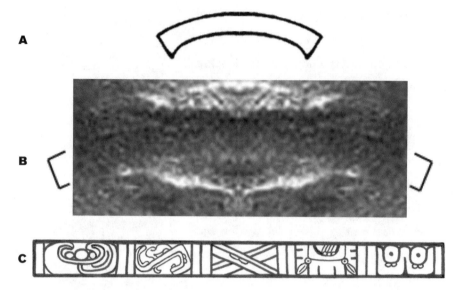

5.13 Comparison of the sky symbols.

A Egyptian Vault of the Sky.

B Mars skyband (pictograph on the humanoid side of the Face).

C Maya skyband (detail).

Drawings by George J. Haas. (Image source: Border of the Lid of Pacal.)

5.14 Aztec sun disk with Tonatiuh at center (detail of Aztec calendar stone).

Drawing by George J. Haas. (Image source: *Gods and Symbols of Ancient Mexico and The Maya*, by Miller and Taube, page 89.)

times portrayed as a creature with a winged, disk-like head; it is set into the lintels of many Maya structures (Figure 5.15).[23]

The central sun disk of Ra is comparable to the Aztec Sun god called Tonatiuh. According to Mesoamerican mythology, the god Tonatiuh presides over the present age of mankind, which is known as the Fifth Sun. His image can be seen in the center of the Aztec sun disk in Figure 5.14.[24] Notice the protruding tongue.

5.15 Maya God Itzam-Ye. Note the winged disk presentation of the bird's head within the spread wings.

Drawing by George J. Haas. (Image source: *The Blood of Kings*, by Schele and Miller, 1986, page 45, figure 22.)

The image of Isis may be seen as the equivalent to the Aztec god Chalchihuitlicue, who is known as Our Lady of the Jade Skirt. She is a Water goddess who is the sister and sometimes wife of the Rain god, who also presides over the Sun.[25] She is outfitted with a long jade dress, water lilies, and plumes of quetzal feathers. The Lady of the Jade Skirt also carries clappers, which are like castanets. They are worn around the wrist and are symbols of water deities.[26] Because of her association with water and the watery, vessel-like environment of the human womb, she also plays an important role as the goddess of birth.[27]

Finally, what we interpret as the hands of Ka are the eyes of Maya First Lord (the humanoid side) in the inverted image. A similar concept is found within a composite image on an Aztec drum, which displays two hands that become eyes (Figure 5.16).

Astonishingly, each of these inverted pictographs found within the humanoid side of the Face relates to the whole concept of the death and resurrection of the Sun in both Egyptian and Mesoamerican mythology.

Because of the oblique angle and narrow presentation of the feline side of the Face, we were unable to complete our analysis of potential pictographs along this side of the demarcation line. We had to wait until a better image of the western side of the Face was obtained to complete our report.

5.16 Aztec drum (stone). Note that the eyes are set within a pair of hands.
Drawing by George J. Haas. (Image source: *Life of the Aztecs in Ancient Mexico,* by Soisson, page 84.)

With the release of the full-faced image of the Face on Mars (NASA/JPL MOC E03-00824—see Chapter 9) on April 8, 2001, our analysis of the feline side of the Face could finally continue. Upon reviewing the new image, we found no evidence of any facial ornamentation on the feline side of the Face that would produce inverted images such as those expressed within the humanoid side.

The symbolism we have found within this amazing structure found at the heart of Cydonia is further evidence of the extraordinary design of this Martian geoglyph. As you will see in the following chapters, the incredible composite design of this geoglyphic mesa is not unique.

Notes

1. Richard C. Hoagland, *The Monuments of Mars: A City on the Edge of Forever,* 4th ed. (Berkeley: North Atlantic Books, 1992), 26–64.

2. Confirmation of the City Square bat is available in MOC image E0500156.

3. J. E. Cirlot, *A Dictionary of Symbols* (New York: Barnes & Noble, 1995), 33.

4. Ibid., 151.

5. Robert Bauval and Adrian Gilbert, *The Orion Mystery: Unlocking the Secrets of the Pyramids* (Toronto: Doubleday, 1994), 116.

6. Ibid., 102.

7. Ibid., 131.

8. Robert Temple, *The Sirius Mystery, New Scientific Evidence of Alien Contact 5000 Years Ago,* 2nd ed. (Rochester: Destiny, 1998), 85.

9. Zecharia Sitchin, *The Wars of Gods and Men, Book III of The Earth Chronicles* (New York: Avon, 1985), 25.

10. Anthony S. Mercatante, *Who's Who in Egyptian Mythology,* 2nd ed. (New York: Barnes & Noble, 1995), 66.

11. Ibid., 81.

12. Robert Bauval and Adrian Gilbert, op. cit., 58.

13. Irene Nicholson, *Mythology of the Americas* (London: Hamlyn, 1970), 155.

14. Ibid., 232. The author offers the reverse side of a sculpture presenting the god Xolotl with the sun tied to his back.

15. Mary Miller and Karl Taube, *The Gods and Symbols of Ancient Mexico and the Maya: An Illustrated Dictionary of Mesoamerican Religion* (New York: Thames & Hudson, 1993), 166, 180. Although some Mayanists suggest that Xolotl represents Venus as the evening star, there is no evidence to substantiate this association. It is more likely that the god of the morning star known as Tlahuizcalpantecuhtli (Lord of the Dawn) also presides over Venus as the evening star, and not Xolotl. At times Tlahuizcalpantecuhtli is represented as a skull-faced god with plumes; he has been identified with Quetzalcoatl. (See Chapter 4, Figure 4.4 for Quetzalcoatl as Venus, back and front views. The skull-faced god on the reverse side of the sculpture may represent Tlahuizcalpantecuhtli.) According to one creation story, Quetzalcoatl is cremated on a funeral pyre and is reborn as Tlahuizcalpantecuhtli.

16. Ibid., 190.

17. Ibid., 80.

18. Ibid., 190.

19. Ibid.

20. The raised foot on tiptoe indicates a dance posture. See Linda Schele and Mary Ellen Miller, *The Blood of Kings: Dynasty and Ritual in Maya Art* (New York: George Braziller, 1985), 275.

21. Robert Temple, op. cit., 85.

22. Linda Schele and Peter Mathews, *The Code of Kings: The Language of Seven Sacred Maya Temples and Tombs* (New York: Quill, 1998), 416.

23. Ibid., 41.

24. Irene Nicholson, op. cit., 148.

25. George C. Vaillant, *Aztecs of Mexico: Origins, Rise and Fall of the Aztec Nation* (Garden City: Doubleday, 1950), 171.

26. Irene Nicholson, op. cit., 240.

27. Mary Miller and Karl Taube, op. cit., 60.

CHAPTER 6

The Drowned Sea

The City's Main Pyramid

Directly to the north of the "bat in the cross-hairs," which occupies the City Square, is a structure that has become known as the City Center Pyramid or the Main Pyramid (Figure 6.2). Although the entire structure was not captured by the MOC, we approximate its length and width to each be about 3 kilometers (2 miles).

With the many early interpretations of *Viking* data, the Main Pyramid was once thought to be an enormous four-sided pyramid based on an Egyptian model. However, the new MOC image shows that this is not the case. It now appears to be five-sided, segmented by five major spines radiating from the top to an almost circular base. The most prominent anomalies on this structure are an almond-shaped crater accompanied by two adjoining rectangular impressions on the northern end of the pyramid (Figure 6.1). These anomalous features were not only noticed by us and many other researchers, but became the main focus of an investigation set forth by Stanley V. McDaniel of the Society for Planetary SETI Research (SPSR).

Early in his study, McDaniel referred to this formation as a "winged ear"[1] and considered it to be one of the most likely features to be of artificial origin. McDaniel was also one of the first to suggest that the bottom

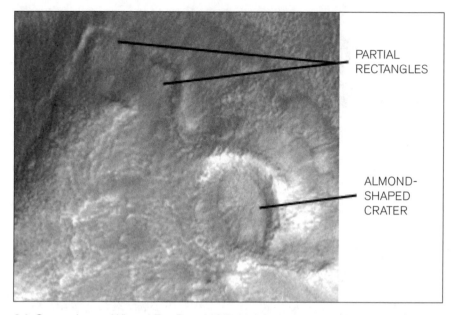

PARTIAL
RECTANGLES

ALMOND-
SHAPED
CRATER

6.1 Context image, Winged Ear. From MOC third Cydonia swath (SP1–25803), showing a portion of the Main Pyramid with partial rectangles and an almond-shaped crater.

of the almond-shaped crater on the Main Pyramid may be coated with "water ice."[2]

The Main Pyramid has a complex series of radiating spines and a set of geometrically shaped features within its surrounding apron. After a considerable amount of time was invested in evaluating these formations, we noted that they were part of a complex set of half-images along the segmented base line (Figure 6.2). We uncovered three pictographic portraits that appear as either the right or left side of a whole image.

When mirrored, the half pictograph marked number 2 in Figure 6.2 forms a striking portrait of a humanoid being along the base line of the Main Pyramid. After studying it, we find it forms a partly human, partly feline, and somewhat amphibious-looking being. He seems to be wearing a uniform adorned with a high broached collar and a large shield-like breast-

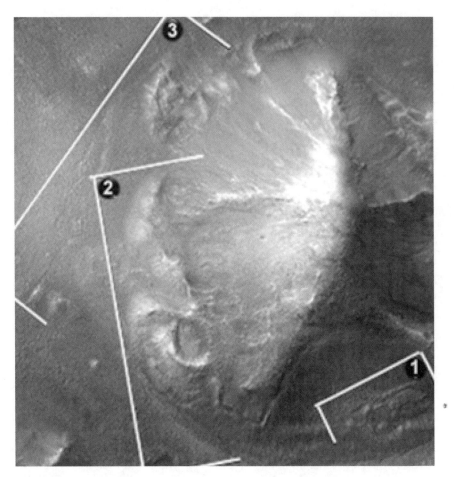

6.2 Context image, Main Pyramid. From the third Cydonia swath (SP1–25803), showing pictograph locations.

1) The Viking (Quetzalcoatl).

2) The Admiral (Enki).

3) Tiger (Water Lily Jaguar).

Additional confirmation of these pictographs is available in MOC image E0201847.

plate, which contains the anomalous winged ear mentioned earlier. This strange character has small squinting eyes, a single tooth, and a broad nose that includes an intricately designed nose ornament. Because of the highly reflective nature of this structure, an aura-like halo appears to be surround-

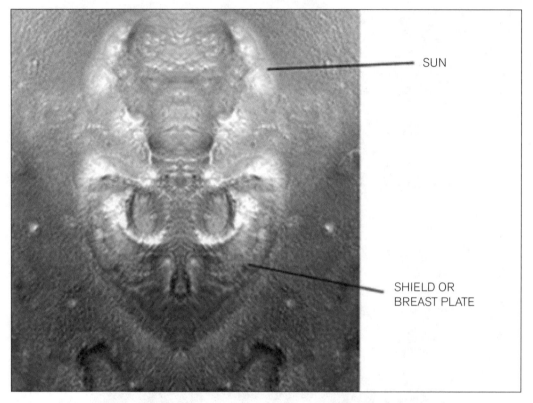

SUN

SHIELD OR
BREAST PLATE

6.3 The Admiral (Enki) (mirrored image of section #2 from Figure 6.2). We believe this image to be the Maya Jaguar god. The image would seem to depict the Jaguar god as the Sun rising from the night sky. Notice the radiant halo around the sides of the head, and the heart-shaped breastplate.

ing his head, which we believe represents the solar aspect of this being. We were so taken by the formal demeanor of this genteel character that we called him "the Admiral," a nickname that would prove very prophetic (Figure 6.3).

After an extensive examination of this image, we eventually decided that it actually represents an aspect of the Maya Jaguar god (GIII) known as Lord Sun (Figure 6.4). The Maya Sun god (Ahau-kin, Lord Sun) is identified most commonly as a human-jaguar hybrid with a gathered lock of hair on his head, a single tooth, and squinting eyes (the result of star-ing at the Sun).[3] There are varying, and at times compli-cated, descriptions available of this god. Linda Schele and Mary Ellen Miller offer the following assessment of Lord Sun as he is depicted in Maya glyphs and art:

> Ahau-kin, or "Lord Sun," has a face with a prominent Roman nose and a square squint-eye. The Kin sign is usually placed ... on the forehead. He has a long, bound hank of hair that falls forward, and his front teeth are often filed into a T-shape.[4]

Figure 6.4 shows a general representation of the Maya Lord Sun as depicted in a jade sculpture. An analytical drawing provided in Figure 6.5 highlights some of the main features of the Martian version of Lord Sun (a.k.a. the Admiral).

6.4 Lord Sun (Maya jade head). Note the bound hank of hair, the single tooth, the broad nose, and the long ears.

Drawing by George J. Haas. (Image source: *Maya Art and Architecture*, by Miller, page 75.)

The Fourth Sun (a Trinity of Gods)

As we explained earlier, the Maya viewed Lord Sun as only one aspect of the many incarnations of the Jaguar god (GIII). Lord Sun was also part of the triad of Creation at Palenque, which included First Father, First Lord, and First Jaguar. Lord Sun was also considered an alternative aspect of First Lord (GI) when identified as Venus and the Sun. This complex trinity of the Cre-ation gods was seen as three separate gods who were ultimately viewed as one and the same being. These three gods appeared in many forms with different visual features and names that occasionally overlapped, causing their specific features to become ambiguous.[5]

In the creation myth, the Creator Couple has four sons: the Red Tez-catlipoca, the Black Tezcatlipoca, Quetzalcoatl, and Huitzilopochtli. Together

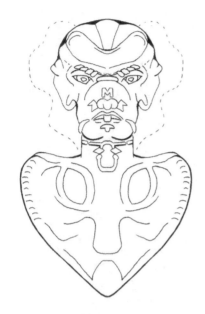

6.5 The Admiral (Maya Jaguar god, Lord Sun).

Note the collar, the bound hank of hair, the single tooth, and the broad nose with its elaborate ornamentation.

Analytical drawing by George J. Haas

these four brothers, who evidently represent the first four Suns of the first four ages in Mayan mythology, also represent the four elements earth, heaven, fire, and sea. The Fourth Sun of this Creation epic is called the Sun of Water.[6] This age is presided over by the Lady of the Jade Skirt, the same god whose image we found inverted on the humanoid side of the Face in Chapter 5 (see the image labeled Isis, Figure 5.6). After the great flood that ends the Fourth Sun or age, the only survivors are fish.[7]

We believe the image in Figure 6.3, which we described above as the Maya Jaguar god in his aspect of the Maya Lord Sun, also represents the Sumerian god Enki, who is the god of water and of creation. Amazingly, noted author and historian Zecharia Sitchin believes Enki also manifested as the Egyptian creator god Ptah. Furthermore, we believe that Enki was ultimately the prototype of the Maya trio of First Father, First Lord, and First Jaguar.

The basic iconography shared by all these gods leads us to suggest that they may have been one and the same. There are two major clues that provide us with this conclusion and may explain the complex threefold aspects of Lord Sun as a humanoid, jaguarian, and amphibious Sun god. The first piece of evidence is presented in Sitchin's book *The Cosmic Code.*

The secret numbers of the gods can serve as clues to the deciphering of secret meanings in other divine names. Thus, when the alphabet was conceived, the letter M—Mem, from Ma'yim, water, paralleled the Egyptian and Akkadian pictorial depictions of water (a pictograph of waves) as well

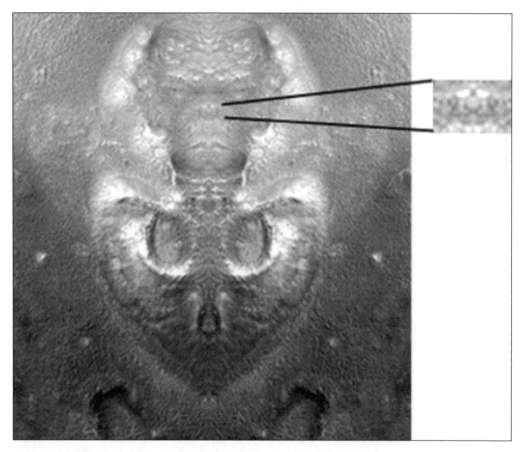

6.6 The Admiral's M-shaped nose ornament (mirrored image of section #2 from Figure 6.2). Note the letter M carries a numerical value of 40. The Admiral image may also represent Enki/First Father/First Lord/First Jaguar (Lord Sun).

as the pronunciation of the term in those languages for water. Was it then just a coincidence that the numerical value of M in the Hebrew alphabet was 40—the secret numerical rank of Ea/Enki, "whose home is water," the prototype Aquarius?[8]

According to Sitchin, this ancient numbering code reveals the numerical placement of the Sumerian gods. The gods are ordered as follows: Anu, the father of the gods, is numbered 60; Enlil, who is Lord of Earth, is num-

6.7A Maya number glyphs. Lord Sun (head variant for number 4). Note the prominent tooth.

Drawing by George J. Haas. (Image source: *Lost Worlds: The Romance of Archaeology*, by White, page 249.)

6.7B Maya number glyphs. Water deity (head variant for number 13).

Drawing by George J. Haas. (Image source: *Lost Worlds: The Romance of Archaeology*, by White, page 249.)

bered 50; Ea/Enki, the Water god, is numbered 40; Sin, the Moon god, is numbered 30; and Shamash, the Sun god, is 20.[9] In examining the portrait of the Martian version of Lord Sun, it is hard not to miss the nose ornament, a very prominent letter M (Figure 6.6). The Maya had an encoding system similar to the Sumerians. In the Mayan lexicon numbers are expressed, not only by the traditional bar-and-dot method, but also by individual portraits of their gods. These numerical profiles, of which there are 20 different heads, are called head variants (Figure 6.7).

The Mayan head variant for the number 13 is seen as a profiled head of a water deity, who is a personification of large bodies of water, such as lakes and oceans (Figure 6.7B).[10] In our modern alphabet the letter M is the thirteenth letter and, as explained by Sitchin, its design is based on an ancient pictograph depicting waves (another water sign).

The Mayan head variant for the number 4, which represents Lord Sun, can be seen in Figure 6.7A. We find it very interesting that this idea of numbered gods was shared by such diverse cultures as the Sumerians and the Maya. In Western culture the number 4 symbolizes the center and the power of the four elements earth, wind, fire, and rain. The number four has a letter companion, the D, which is represented in the original Greek form as the delta, a triangle. This triangle symbolizes a doorway[11] or, because of its shape, it could also be seen as a tetrahedron. This idea of the Sun god's connection with the tetrahedron will be examined further in Chapter 7.

On a Sumerian cylinder seal, the god Enki is depicted with fish flowing out of his arms while standing over a small stag (Figure 6.8). Enki's original name was Ea, which means "whose house is water."[12] In Sumerian mythology, Enki is the Water god and the Creator of civilization. The Sumerians believed that a freshwater ocean that lay beneath the Earth, called the *abzu*, was his underwater realm. This subterranean ocean fed the springs, wells, streams, lakes, and rivers of the world.[13] His watery realm was inhabited by many sea creatures, all of which served him—including the turtle.[14]

6.8 The Water god Enki with fish flowing from his arms (Sumerian cylinder seal, circa 2250 B.C.). Note the two-faced god Isimud on the right.

Drawing by George J. Haas. (Image source: *Mythology: An Illustrated Encyclopedia*, by Cavendish, page 87.)

According to Sitchin, it is in Enki's honor that two of the constellations of the zodiac are named the Waterman (Aquarius) and the Fishes (Pisces).[15] The priests who oversaw Enki's worship were dressed as hybrid fish-men. These fish-garbed figures, referred to as the popular guardian known as Oannes, were bearded humans who wore a fish head pulled over their forehead; the full body of the fish hung down their back (Figure 6.9A).[16] They held crescent-shaped "buckets," symbolizing the Moon, which may have held water.

6.9A Fish-garbed headdress. Oannes (relief from Temple of Ninurta, Kalhu). Note the fish hanging down the back and the crescent-shaped bucket.

Drawing by George J. Haas. (Image source: after Layard, *Gods, Demons and Symbols of Ancient Mesopotamia*, by Black and Green, page 83.)

6.9B Fish-garbed headdress. Maya figure (Hero Twin?). Note the fish hanging down the back.

Drawing by George J. Haas. (Image source: Tro-Cortes Codex)

Although an ocean away from the cultures of Mesopotamia, a comparable fish-men motif is also found in the New World. In Mesoamerica the Maya depicted a similar fish-garbed figure in the Tro-Cortesianus Codex (Figure 6.9B). Here a figure on stilts wears a fish head pulled over his forehead as the body of the fish hangs down his back just as we see with the Sumerian figure known as Oannes.

The Maya Sun god who ruled the Fourth Sun, called the Sun of Water, is seen here as Enki's contemporary force. The dual relationship stems from the fact that the heat of the Sun causes droughts, which dry up lakes and sometimes entire oceans. In Maya myths, the stag is seen as a sign of drought, which is caused by the Sun,[17] while in Sumerian myths Enki is sometimes called the Stag of Abzu.[18] Notice the subdued stag reclining beneath Enki in the cylinder seal in Figure 6.8. There will be more about this stag and its relationship with the original Face on Mars in Chapter 11.

The Viking (a Bearded Quetzalcoatl)

The pictograph from the Main Pyramid marked number 1 in Figure 6.2 consists of a half face which, when mirrored, is reminiscent of a typical Viking lord (Figure 6.10). The head appears to be composed of a full-bearded face with braids, eyes, nose, and mouth. The head is complete with a Viking-style helmet. The portrait is perfectly framed and we found its design simply amazing. The Viking head is roughly 700–800 meters (half a mile) in length from the top of the helmet to the neck.

Although our initial judgment of this structure led us to believe that it represented a Viking-like portrait, we acknowledge that this image also bears many features similar to the Aztec depiction of Quetzalcoatl (the Feathered Serpent), as seen in Figures 6.11 and 6.12. This image of Quetzalcoatl is bearded and wears a helmet strikingly similar to the Mars figure. The portrait includes a lower-helmet extension and long braids made of serpents. If one

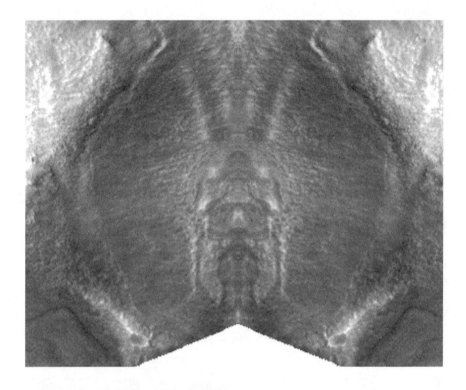

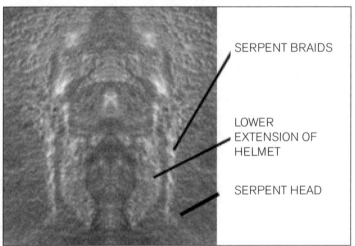

SERPENT BRAIDS

LOWER
EXTENSION OF
HELMET

SERPENT HEAD

6.10 The Viking (Quetzalcoatl) (mirrored image of section
#1 from Figure 6.2). Note the perfect framing of the
bust (top) and the helmet, beard, and braids in the enlarged
image (bottom).

6.11 Winged Quetzalcoatl (Aztec Codex). Note the wings and the vessel
he carries in his left hand.

Drawing by George J. Haas. (Image source: *Mysteries of the Mexican Pyramids,*
by Tompkins, page 385.)

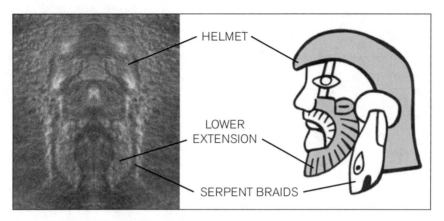

HELMET

LOWER
EXTENSION

SERPENT BRAIDS

6.12 Comparison of The Viking and Quetzalcoatl.

RIGHT, Detail of Martian Viking.

LEFT, Detail of Aztec Quetzalcoatl.

Note that the lower extension of Quetzalcoatl's helmet and the serpent braids match the Mars image.

looks closely at the braids of the Martian image in Figure 6.12, it also appears as though they are formed by a coiled serpent.

The Assyrians, who emerged in Mesopotamia around 2000 B.C., assumed many aspects of Sumerian culture. In comparing the images in Figures 6.11 and 6.13, one can't help but notice that both wear wings, have beards and braided hair, wear helmets, and carry a small basket in their left hands. These small baskets are also a common motif in Maya and Olmec imagery.

Traditionally, Quetzalcoatl was described as a bearded, fair-skinned, helmeted, god-like king. Some researchers believe he may have been of Scandinavian or Semitic origin. According to Aztec mythology, Quetzalcoatl bestowed the knowledge of the arts and sciences on their people. Then he left, sailing east across the sea on a "raft of serpents," promising he would return in the year "1 Reed." Because of the use of serpent motifs on the bows of their ships, this Aztec reference to a raft of serpents has been used as evidence to further the idea of early Viking contact in the Americas.

The Maya year 1 Reed occurred in fifty-two year cycles and it was in precisely one of these years, 1519, that a helmeted, bearded explorer named Hernando Cortez arrived in the Americas. Because of the precise timing of his arrival and his divine appearance, Cortez was tragically mistaken by Montezuma to be the returning Quetzalcoatl.

Quetzalcoatl, as he was called by the Aztecs, or Kukulkan, as he was known to the Maya (both mean the Feathered or Winged Serpent), was the dual embodiment of Venus and the Sun. Zecharia Sitchin believes that Quetzalcoatl was actually Ningishzidda, the youngest son of the Sumerian god Enki. Sitchin also believes that Ningishzidda/Quetzalcoatl was also the Egyptian god Thoth, son of Ptah, who came to the Americas with his African followers (the Olmec).[19] It is interesting to find that a portrait of Enki (as the Water/Sun god) lies right next to Quetzalcoatl (seen as a seafarer) at the Main Pyramid. Both of these characters play an important role in this overall story, as does the third image, the squatting feline.

6.13 Assyrian winged deity (Genie). Note he carries a vessel of water in his left hand.

Drawing by George J. Haas. (Image source: *Larousse Encyclopedia of Mythology*, by Aldrington and Ames, page 69.)

The Water Lily Jaguar

The third portrait along the apron of the Main Pyramid is centered immediately above Enki in Figure 6.2 and is marked by the number 3. This half image when mirrored becomes that of a feline believed to be a tiger in a sitting position (Figure 6.14B). When the Mars tiger is compared to an Asian tiger, the resemblance between the two cats is very impressive (Figure 6.14A). Notice the zigzag stripes on the face of the Cydonia tiger, which measures roughly two and a half kilometers from head to toe (one and a half miles).

The first oddity we noted in this pictograph was that the squatting feline at Cydonia wears a strange "tie" or "scarf" hanging from its neck (Figure 6.17). It also has a headdress and a feature resembling a headband that sweeps across its forehead (Figure 6.16). Crowns and headdresses were depicted on both humans and animals throughout Mesoamerican art and were cultural markers of divinity. Between the ears of the sitting feline is an odd floral feature, which could be interpreted as a "water-lily" ornament. This particular ornament reminded us of the Maya Water Lily Jaguar.

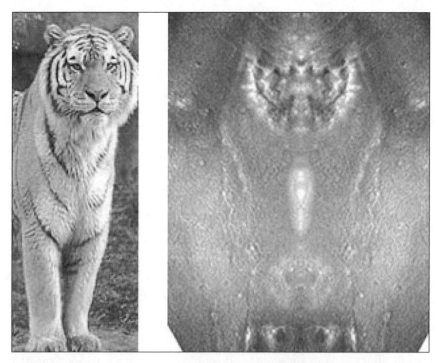

6.14A Photograph of an Asian tiger.

6.14B The Martian tiger (Water Lily Jaguar) (mirrored image of section #3, Figure 6.2).

In Maya iconography, the normal attire of the Water Lily Jaguar (another form of the Jaguar god) is a knotted scarf worn around his neck and a water-lily sign featured on his brow or atop his head (Figure 6.15).[20] The water-lily sign, a metaphor for royal power, can take the form of a leaf, a flower, or a net-like pattern as seen on the shell of turtles.[21] The water-lily blossom sign is also thought to further represent the jaguar's love of the water.[22]

6.15 Water Lily Jaguar (seated within a water lily). Note the scarf and the water lily symbol on his head.

Drawing by George J. Haas (Image source: *Study of Maya Art*, by Spinden, page 135.)

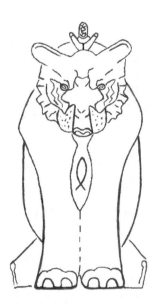

6.16 The Martian Tiger (Water Lily Jaguar) (analytical drawing of mirrored Martian Tiger). Note the stripes on the face, the scarf, and the water lily symbol between his ears.

Drawing by George J. Haas

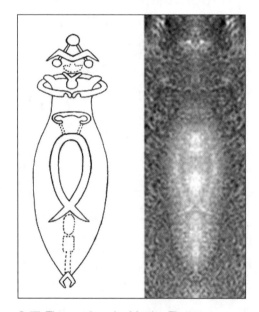

6.17 The scarf on the Martian Tiger.
LEFT, Tiger's scarf.

Analytical drawing by George J. Haas

RIGHT, Tiger's scarf (detail of Figure 6.14B). Note the fish symbol.

Upon closer examination, the scarf feature of the Cydonia tiger has within its amphora design the image of an open-tailed fish, much like the common Pisces symbol of the zodiac and the one used by Christians to symbolize the resurrected Christ. In the eighteenth Sura of the Muslim holy book, the Koran, the fish is seen as a symbol of resurrection.

An ancient Zapotec urn, found in Monte Alban, Mexico, is sculpted in the shape of a squatting, snarling jaguar very much like the Cydonia tiger (Figure 6.18).[23] The most interesting feature of this statue is the knotted scarf that hangs around its neck. In Mesoamerican culture, the scarf was a symbol of resurrection.

In comparing the Cydonia tiger and the Zapotec jaguar, they both wear symbols of resurrection on their chests and are presented in a seated position with their front legs bowed inwards. The Cydonia tiger has a headband bearing a cat face, while the Zapotec jaguar bears a scarf with a cat face within the knot (Figure 6.19).

In reviewing this amazing overlapping symbolism, we assert the central figure on the Main Pyramid to be the Sumerian god Enki (or the Water Sun, who is the Maya Jaguar god as the daytime Sun). Beside him we have identified the Maya god Quetzalcoatl (or the Sumerian god Ningishzidda, Enki's son in his guise as the morning star, Venus). And directly above them we have a third aspect of the Maya god as the Water Lily Jaguar (seen as the nighttime Sun or the Sun of the Underworld). The nighttime Sun, of course, is the Moon. The Jaguar god was also known as Balam-U-Xib or Jaguar Moon Lord.[24] The tiger was also an ancient Chinese symbol of the New Moon[25] (a sign of resurrection). The inherent associations between these three characters are significantly entwined, and to have them all depicted on the same structure on Mars is incredible!

These three pictographs combined cover over half of the apron around the Main Pyramid. Unfortunately, the Mars Orbital Camera did not cap-

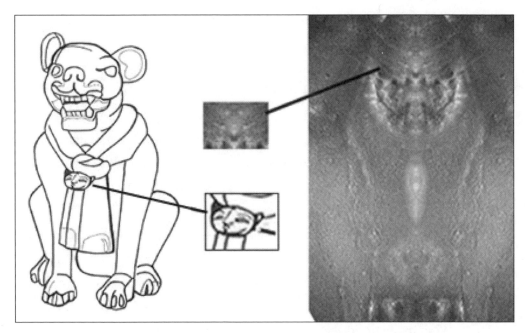

6.18 Mesoamerican Crouching-Jaguar (Urn), 200 B.C. Note the scarf.

Drawing by George J. Haas. (Image source: *Great Ages of Man*, by Leonard and editors of Time-Life Books, page 3.)

6.19 Martian Tiger (mirrored image of section #3, Figure 6.2). Note the scarf feature and the crouching posture of the tiger.

ture the remaining portion of the Main Pyramid and we are unable to confirm the existence of any other related pictographs within its surrounding apron.[26]

The Jaguar Protector

As we have learned, the Jaguar god is a key figure in Mesoamerican religion and culture, manifesting in many entirely different personas. He is called Jaguar of the Underworld, Jaguar Sun (Lord Sun), Jaguar Night Sun, Water Lily Jaguar, and many other names. Snarling jaguar heads appear atop the acropolis in Cerros and throughout the Maya and Olmec world. The jaguar was essentially an Uay, a divine manifestation of a human being, or a god in animal form.[27]

A carved lintel panel from a temple at Tikal (Lintel 3 from Temple I) depicts King Hasaw-Ka'an being carried on a palanquin. He wears a Sun god (Lord Sun) headdress and above him is the snarling "Jaguar god Protector" (GIII as the Jaguar god again), slashing his claws into the Otherworld.[28] On the arm of the Jaguar god Protector is a wristband with a split mask of a face and a bird. The wing of the bird is composed of the fanned-out, slashing claws of the Jaguar Protector (Figure 6.20).

A mirrored image of the lintel done by a split between the half face and the half bird on the Jaguar Protector's wristband proves fascinating (Figure 6.21). The half image of the face on the left side of the wristband becomes a glyph, which may represent the Jaguar god. The half-bird image on the right side becomes an open-winged horned owl.

Immediately to the south and east of the Main Pyramid of the city complex at Cydonia is another multifaceted structure. The top left side of this structure portrays another feline portrait. The tight crop of this snarling cat's head looks a lot like the cougar logo used by the Lincoln-Mercury Car Company for its Cougar line of sporty cars (Figure 6.22A). We believe this amazing profile actually forms a snarling feline head that takes on the persona of the Jaguar Protector seen on the Tikal lintel (Figure 6.20).

Realizing that this structure could also have been meant to be mirrored, a mirror image was created along a vertical line, marked by small mounds. The result creates a face very similar to the mirrored glyph on the Tikal lintel between two snarling Jaguar Protectors (Figure 6.23). The resulting image of the jaguars is approximately 1 kilometer from head to toe (roughly three-quarters of a mile).

Figure 6.24 is a comparison of the mirrored Jaguar Protector in the Tikal temple lintel in Figure 6.21 and the Mars image from Figure 6.23. Both

6.20 Jaguar Protector (detail of wooden door lintel from Temple I, Tikal).

Drawing by George J. Haas. (Image source: After John Montgomery.)

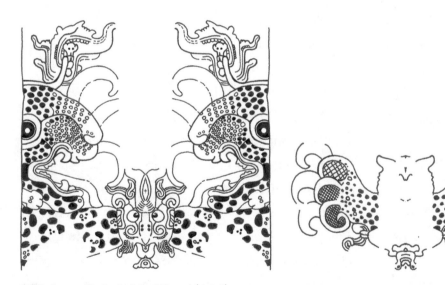

6.21 Jaguar Protector's wristband (detail).

LEFT, Left half of wristband, mirrored (face or mask). Note how the two Jaguar Protectors flank the mask.

RIGHT, Right half of wristband, mirrored (spread-winged bird). Note how the Jaguar Protector's two paws form wings.

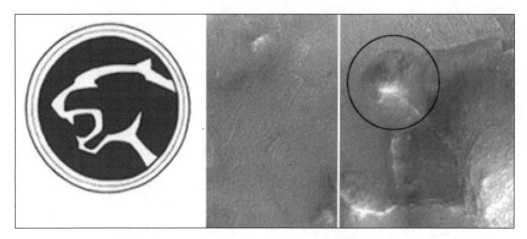

6.22 Snarling jaguar.

LEFT, Cougar logo (Mercury automobiles).

Drawing by George J. Haas

RIGHT, Context image for Snarling Jaguar (third Cydonia swath, SP1–25803). Note that the demarcation line marks the spot where the jaguar head was mirrored to create Figure 6.23.

form similar faces in the middle. The face in the Mars image wears trumpet-shaped ear flares. Conch-shell trumpets were worn as ear ornaments by the Maya. By sounding the conch-shell trumpets, shamans were able to open portals to the Otherworld.[29]

According to the Maya shamans, the Jaguar Protectors, who were also called the Balamob, were seen as sentinels. These jaguars were responsible for protecting the towns and fields of the kingdom in conjunction with the Sun god, Lord Sun.[30] The Balamob or Jaguar Protectors, like the four sky-bearers and rain gods, were all fourfold beings associated with the cosmos and the four cardinal points. These Jaguar Protectors, however, were seen as slightly different divinities in that they operated at a level that was more "intuitive to the human landscape."[31]

With this in mind, it only seems fitting that an image of a Jaguar Protector would be found just off the edge of the Main Pyramid, protecting the

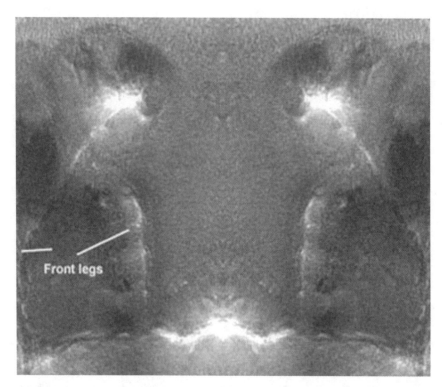

6.23 Martian Jaguar Protectors. Note the mirrored snarling Jaguar Protectors with jaguar mask between them at bottom. The jaguar mask appears to hold up the front legs of the two Jaguar Protectors.

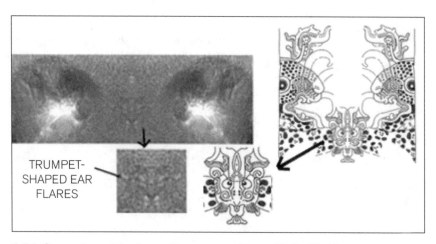

6.24 Comparison of the Jaguar Protectors on Mars with the Tikal lintel.

Fourth Sun of Creation, Lord Sun. Ultimately, what we found in the next level of our research was that this jaguar was protecting more than just the Fourth Sun. It was pointing the way to another being, like the Jaguar Protectors, associated with the cosmos and the four cardinal points—a sky-bearer.

The Cosmic Turtle

The turtle is seen as a symbol of the cosmos in many cultures. This primordial model is supported in the simple construction of its outer shell, which consists of a combination of round and square forms. The Heavens are represented in the rounded top half of the turtle's shell (the carapace), the Earth in the square, flat bottom portion of the shell (the plasteron). Its four short, bulky legs act as pillars enabling the turtle to bear the weight of the cosmos.[32]

The Cosmic Turtle in Maya mythology floated in the primordial sea of creation, and it was from a small crack (or cleft) in its back that the First Father was reborn after being resurrected by his twin sons. As soon as First Father was reborn, a triangular hearth of three "throne stones" was set up by the gods on August 13, 3113 B.C.[33] This triangle symbolized the center of the new order (the shape of the Greek letter D, delta, is a triangle and signifies a door or a portal). The hearthstones were also seen as the three stars of Orion's Belt, which were set across the back of the Cosmic Turtle.

The first of the stones was set in the form of the jaguar throne.[34] This jaguar throne acted much like the Jaguar Protector, guarding the cosmic center like a sentinel. Maya mythology states that 542 days after the resurrection of First Father through the cleft in the Cosmic Turtle, four sky-bearers set up the four sides and corners of the cosmos. They then erected the center tree, and together these events signified the beginning of the Fourth Creation, or Fourth Sun.[35]

The image in Figure 6.25 is from a Maya stela that shows the Cosmic Turtle with a water-lily stem and bulb protruding from its shell. Since the water lily floated on the surface of the water, it was also a symbol for the division of the Middleworld and the Underworld from where the First Father was resurrected.[36]

On the same structure that displays the portrait of the Jaguar Protector (another aspect of the first of the three hearthstones), we believe we have found just such a Cosmic Turtle, complete with a water-lily bulb. We have estimated that the overall shape of the structure that displays the profiled head of the Jaguar Protector is consistent with the shape of a turtle. Although the whole turtle structure was not captured by the Mars Orbital Camera in this third

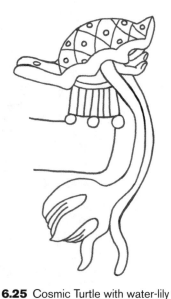

6.25 Cosmic Turtle with water-lily bulb (Madrid stela, detail).

Drawing by George J. Haas. (Image source: *Study of Maya Art*, Spinden, page 18.)

swath, we added the remaining data from the April 22, 2000 release of the right side (M09–05394) and created our own composite image (Figure 6.26).

Once the image is completed, the sea turtle features become even more pronounced. A turtle flipper can be seen on the left side of the domed shell that exhibits the pronounced corrugated turtle-shell patterns around the top rim. A feature resembling the water-lily stem and bulb from the Madrid stela can be seen at the bottom left of the shell. The bulb of the water-lily plant is formed by the crater-like structure at the bottom of the stem.

If a mirror split is performed down through the center of the face and continued along the spine of the turtle, the left half of the profile becomes a human-like face. The overall body of the structure forms a turtle shell with two flippers and a set of water-lily bulbs growing from the bottom of the turtle (Figure 6.27).

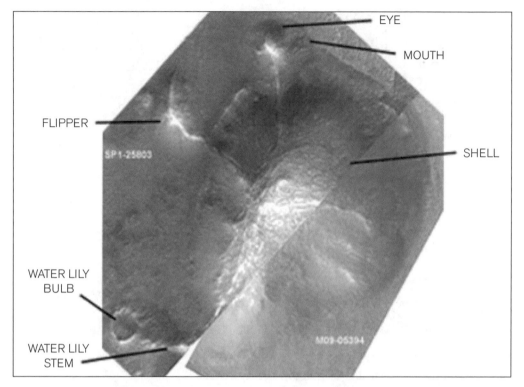

EYE

MOUTH

FLIPPER

SHELL

SP1-25803

WATER LILY
BULB

WATER LILY
STEM

M09-05394

6.26 The Cosmic Turtle mesa (composite image of third *MGS* swath, MOC SP1–25803 and MOC M09–05394). Note this composite image is reminiscent of the Rabbit/Duck image in Chapter 3 (Figure 3.12), in which the image forms two different faces when viewed from opposite directions. In this case, a jaguar head faces left and a turtle head faces right.

The head rising out of the cleft of the turtle on Mars is believed to represent the ancient Maya god of "Sky Bearers" called Pawahtun (Figure 6.29). This mature head has a remarkable resemblance to a face seen on a Maya sculpture of a Pawahtun in the guise of a turtle (Figure 6.28). Notice the common facial features and disk-shaped ear flares of both the Maya and Martian heads as they emerge from a turtle carapace. An analytical drawing of the left side of the Cosmic Turtle is provided in Figure 6.30.

Located on one of the outer piers of the Lower Temple of the Jaguar at Chichen Itza is one of the most revealing images of Maya cosmology. Fig-

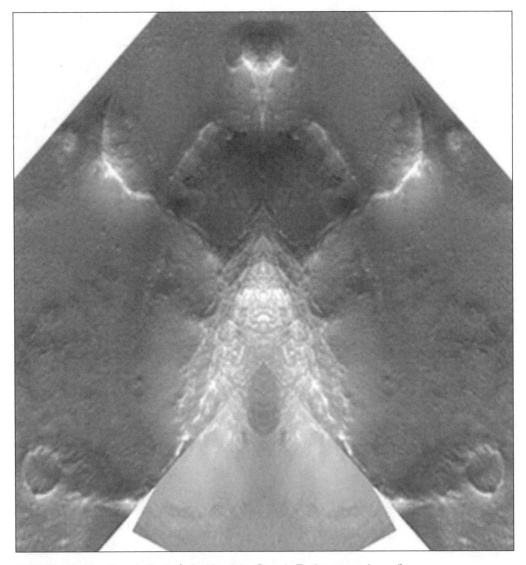

6.27 The Martian Cosmic Turtle (left side of the Cosmic Turtle mesa, mirrored).
Note the head rising out of the cleft, the flapper, and the water-lily stem with bulb.

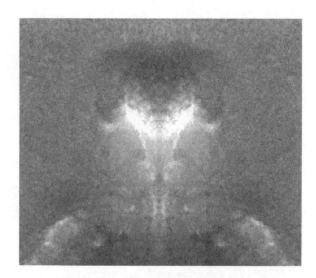

6.28 Pawahtun head with turtle shell (Maya, clay/paint). Note the shape of the detachable head and disk-shaped ear flares.

Drawing by George J. Haas. (Image source: Justin Kerr K2980b.)

6.29 Head of Pawahtun emerging from turtle shell (left side of Jaguar/Turtle head from Figure 6.27 mirrored). Note the mustache/nose plug, and disk-shaped ear flares.

HEAD (DETAIL)

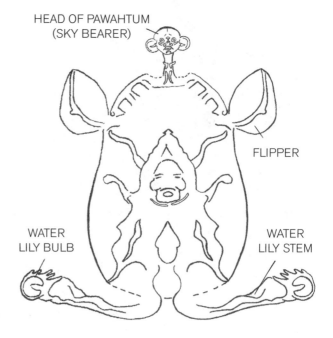

HEAD OF PAWAHTUM
(SKY BEARER)

FLIPPER

WATER
LILY BULB

WATER
LILY STEM

6.30 Cosmic Turtle with water lily and head of Pawahtun (left side of Cosmic Turtle mesa, mirrored, with head detail).

Analytical drawing by George J. Haas

ure 6.31A depicts Pawahtun, one of the "old gods" who held up the sky at the time of Creation.[37] The facade features an image of Pawahtun standing with a turtle shell fastened to his chest. The turtle shell suggests that this old god is to be seen here as a personification of the Cosmic Turtle that floated on the surface of the primeval waters at the time of Creation. The Pawahtun god also wears ear flares and a medallion with a "zero" sign (Figure 6.31B), again signifying the beginning of time.

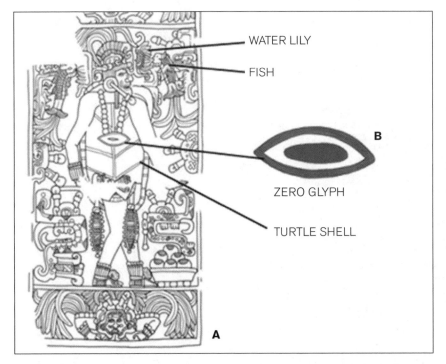

WATER LILY

FISH

B

ZERO GLYPH

TURTLE SHELL

A

6.31A Pawahtun personifying the Cosmic Turtle (outer facade of the Lower Temple of the Jaguar at Chichen Itza). Pawahtun (at the day of Creation). Note that a turtle shell forms the torso of Pawahtun and he wears round ear flares and a medallion with the Maya glyph for zero.

Drawing by Linda Schele, © David Schele. (Courtesy of Foundation for the Advancement of Mesoamerican Studies,Inc., www.famsi.org.)

6.31B Maya zero glyph (found in Maya codices).

Drawing by George J. Haas. (Image source: *Introduction to the Study of the Maya Hieroglyphics,* by Morley, page 92.)

A lily blossom can be seen hanging down in front of Pawahtun's head-dress with a fish nibbling at its petals. A lily blossom and a fish are also featured on the Cydonia landscape.

When the right side of the Mars turtle is mirrored, a turtle's head is seen withdrawing into its shell (Figure 6.32). This is in opposition to the left split, where the head and neck are noticeably protruding. Surprisingly, a floral pattern emerges from the center of the shell which forms a fully opened water-lily blossom.

In Hinduism the symbol of a withdrawn turtle suggests "involution and return to the primeval state and therefore of a basic spiritual attitude."[38] This idea of the water lily and its connection to a spiritual birth and res-

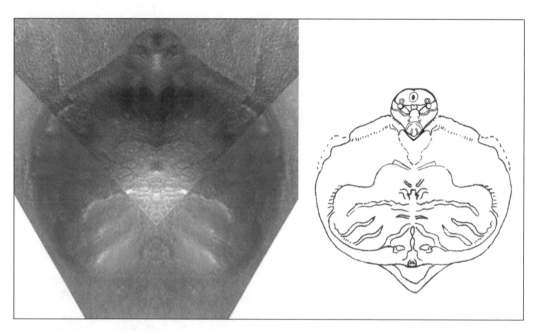

6.32 Withdrawn turtle with a water-lily blossom on its shell (right side of the Cosmic Turtle mesa, mirrored). Note that the turtle has retracted into its shell while his head pokes out. A large water-lily blossom decorates the back of the shell.

6.33 Withdrawn turtle with a water-lily blossom on its shell. Note the turtle's head at top.

Analytical drawing by George J. Haas

urrection is also found within the Hindu religion in the form of the lotus. An analytical drawing of the right side of the withdrawn turtle with a water-lily blossom on its shell is provided in Figure 6.33; it highlights some of the details.

Again, as with the Main Pyramid trio of gods, we are presented with a structure that has a threefold aspect in Maya mythology. Not only is the profiled jaguar head on this structure a portrait of Jaguar Protector, but when mirrored, it transforms into the head of an aged sky-bearer protruding out of a turtle's shell.

If we go back and look at the significance of the beginning of the Fourth Sun, we find one of its results was a Great Flood, in which the only survivors were fish. You will note that there are twin fish swimming around the head of the Pawahtun on the facade at Chitchin Itza (Figure 6.31A). Could this be another clue to the strange shapes of sea creatures that surround the Main Pyramid? To the north and south of the Main Pyramid are megalithic structures designed to look like sea creatures. These creatures include the aforementioned turtle and a dolphin, as well as a few fish!

The Headless Fish

At the far south end of the third Cydonia swath there is a large structure that looks very much like a headless fish, partially filleted or cut up (Figure 6.34). The appearance of the fish's body suggests that it has been cut open, mangled, and left with its spine exposed. The tail appears to be missing and the decapitated head lies right next to the fish's body with its mouth gaping open.

The fish plays a major role in the mythology of the Maya; in the *Popol Vuh* it is used as a reference to the resurrection of the original Hero Twins. In the story, after the Hero Twins sacrifice themselves in an oven, their

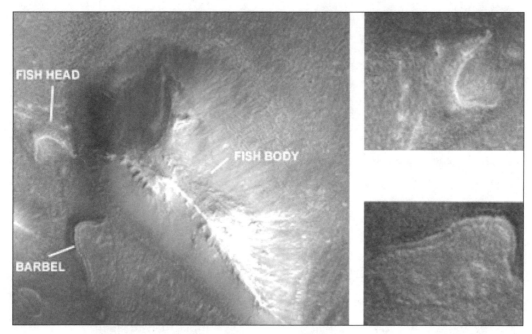

6.34 Headless fish (third *MGS* swath, MOC SP1–25803).
LEFT, Headless fish. Note the inverted fish head and the barbel.
RIGHT, Detail of fish features. Top–fish head. Bottom–barbel.

bones are cut up and thrown into a river. After only a few days they reappear, transformed into catfish. Figure 6.35 provides a detail of one of these transformed catfish. Notice how the tail fin is bifurcated, echoing the shape of human legs. The *Popol Vuh* describes this resurrection and transformation of the Hero Twins into catfish as follows:

> ... and on the fifth day they reappeared. They were seen in the water by the people. The two of them looked like channel catfish when their faces were seen by Xibalba....[39]

The Headless Fish structure at Cydonia (Figure 6.34) seems to duplicate this episode in Maya mythology with amazing accuracy. The decapitation sacrifice and resurrection of the first set of Hero Twins (First Lord and First Jaguar) as catfish is displayed here in stunning detail. The Headless Fish has a curving, L-shaped appendage branching off the lower gill area, which is most interesting. This anomalous feature has been discussed by other researchers as resembling an enormous wall. When viewed in the context of its surrounding aquatic companion, this wall structure looks more like a fish barbel, or bristle. Among fish species, the barbel is commonly recognized as a feature of the catfish.

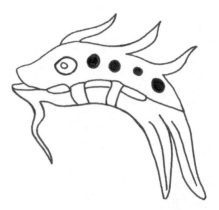

6.35 Hero Twin as a catfish (detail of Maya cylinder vase). Note the barbel and the leg-shaped tail fin.

Drawing by George J. Haas. (Image source: *Painting the Maya Universe*, by Reents-Budet, page 242.)

The Headless Fish is missing not only its head but also its tail. At the front of the fish body, where the head would be, two small parallel lines can be seen. We believe these two lines are intentionally placed markers, which provide a demarcation for mirroring (Figure 6.36). If a mirror split is done between the lines and continued through the body, magically the fish is resurrected. A fish's head is formed when one side is mirrored, while the missing tail can be found in the opposite split (Figure 6.37). Amazingly the surrounding rock formation gives the appearance of the fish head underwater, swimming right at you, as if in a channel. Notice the small, cat-like "whiskers" above the fish's mouth and the teeth.

The Headless Fish that lies just south of the Main Pyramid also parallels the significance of the fish sign found in the scarf of the tiger, which we saw around the apron of the Main Pyramid. But this repetitive message of duality, fish, water, and death and resurrection doesn't end here. Our next revelation begins with yet another aquatic creature, found just to the north of the Main Pyramid, in the form of a dolphin.

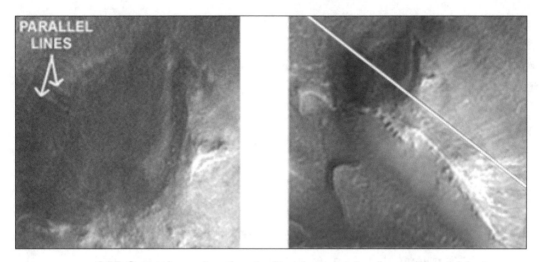

6.36 Context image. Location of split performed on headless, tail-less fish body (third Cydonia swath, MOC SP1–25803).

LEFT, Parallel marker lines.

RIGHT, Demarcation line of the split. Note that although the SP1–25803 swath ends where the fish tail would be, the 1976 *Viking* image of the same area shows that the structure does not continue.

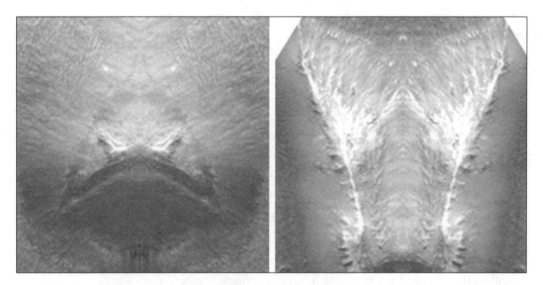

6.37 Swimming fish and fish tail.

LEFT, Swimming fish (mirrored image of fish head in water).

RIGHT, Fish tail (mirrored image of fish tail).

The Dolphins

Between the Main Pyramid's base line and an adjacent structure known as the "Fort"[40] is a large pictograph in the profiled shape of a dolphin (Figure 6.38). The contoured shape of the dolphin appears to be cut right into the surface of Mars in a manner similar to the Nazca line drawings in Peru. The imprint of the dolphin includes a prominent dorsal fin, a flipper, a crescent-shaped tail, and a bottle-shaped nose. The overall contour of this dolphin is uncanny. A silhouetted overlay is provided over the dolphin pictograph for comparison (Figure 6.38).

To further increase our curiosity concerning the apparent connection of this Martian structure to Mesoamerican culture, we were not surprised to find that this docile marine mammal was regarded as a sacred emblem among the Olmec. In Veracruz an axe, or hacha, with an Olmec-like head was found bearing a stylized dolphin, worn as a headdress (Figure 6.39). These hachas or axe heads were used as trophy heads that were displayed at ball courts throughout ancient Mesoamerica.[41]

6.38 The Main Pyramid dolphin (third Cydonia swath MOC SP1–25803).
LEFT, Detailed clip of original dolphin pictograph.
RIGHT, Highlighted silhouette of dolphin pictograph.

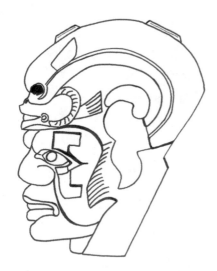

6.39 Dolphin headdress (votive hacha, Central Veracruz, Classic period). Note this stone axe trophy head, with stylized dolphin headdress, is classified as a dolphin by archaeologists and not the authors.

Drawing by George J. Haas. (Image source: *Mexican National Museum of Anthropology*, by Bernal, page 127.)

Although not much is known about the significance of this dolphin motif, one distinction should be made. Although dophins may look like fish, they actually fall within a class of aquatic mammals. The Olmec may have found this transformational aspect of the dolphin appealing because of its tendency to leap out of the water, giving the appearance of living between two worlds where the "sea and the sky meet."[42] This dual identity of the dolphin may have been one of the reasons the Olmec incorporated its shape into their ball court iconography. The ball game's courtyard was seen as a stage for the re-enactment of Creation, and the dolphin may also be seen as a symbol of that transformation and Creation.

The presence of this dolphin gives continuity to the aquatic theme surrounding the image of the Admiral on the Main Pyramid. Indeed, the equating of this image with Enki, "whose house is water," could not be more fitting. The image in Figure 6.40 shows the location of the sea creatures in relation to the image of the Admiral. We have isolated these structures in Figure 6.41 by artistically creating an ocean that perhaps was once present where there is now Martian desert.

The immense dolphin pictograph is not the only dolphin image to be found in the Cydonia complex on Mars. In the first Cydonia swath (MOC SP1–22003), taken in April 1998, there is another dolphin-shaped structure located just below the Animal mesa that features the split faces of 18 Rabbit and Smoke Monkey. This formation, which resembles the profile of a swimming dolphin surfacing in the water, has been called the "Coat

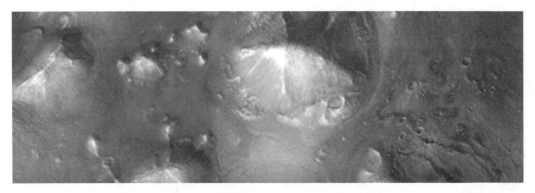

6.40 Context image of the Main Pyramid with surrounding sea creatures (third Cydonia swath, MOC SP1–25803).

LOCATION OF
"THE ADMIRAL" (ENKI)

6.41 Location of sea creatures relative to the Admiral (Enki), on Gray Field (artistic interpretation of aquatic theme found in Figure 6.40).

Clockwise from top left: Headless Fish, Open-Mouthed Fish (possibly a Sunny), the Main Pyramid, Dolphin, and Turtle.

Hanger" by other researchers. Our identification has been fostered by the dorsal fin and the unusual linear configurations that are found below the dolphin's body, creating the lower bar of a coat hanger (Figure 6.42).

These striations are in an almost ruler-straight arrangement of needle-like white lines that produce a high albedo effect. Similar linear markings were also found on the collar feature of 18 Rabbit discussed in Chapter 2

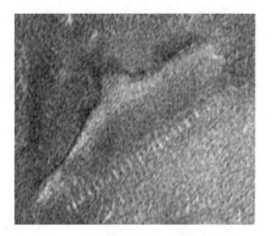

6.42 Coat-hanger dolphin (first Cydonia swath, MOC SP1−22003). Note the arrangement of needle-like white lines under the swimming dolphin.

6.43 Coat-hanger dolphin. Note the bottle nose, dorsal fin, tail, and eye.

Analytical drawing by George J. Haas

(Figure 2.6). The overall shape of the "coat-hanger dolphin" indicates a prominent dorsal fin, a downcast tail, and a distinct bottle-nose (Figure 6.43). There is also a hint of an eyeball in the head area. Remarkably, all these features appear to be within the right place and within the proper orientation to form an almost perfect dolphin.

In Greek mythology, the dolphin's divine image was found beside Apollo's tripod (another triangle) at the oracle of Delphi. Apollo, the Sun god, took the form of a dolphin to reach the shore of Crisa on his way to Delphi.[43] The word *Delphi* is derived from the word *delta* (the Greek letter D) which, as we have seen, means door and is triangular in shape.

The Main Pyramid dolphin lies right next to the pictographic portrait of the Admiral, whom we have identified as the Maya Sun god Lord Sun, whose head variant number is 4. Again we find the overlapping symbolism of the Sun and the number 4, which is also the Greek letter D (or delta, which brings us back to a triangle). Richard Hoagland has also noted that the underlying structure on which the coat-hanger dolphin rests is triangular or tetrahedral in shape.[44]

In Greek mythology the idea of gods being transformed into dolphins was carried over to mortal sailors who were lost at sea. This myth is illustrated on the inside bowl of an ancient Greek drinking cup (Figure 6.44).

The cup depicts the story of Dionysus, the god of wine (the "blood of the gods"). On his way to Naxos, pirates abduct him and hold him prisoner. In an effort to escape, he causes grapevines to fill the entire ship and the pirates become so frightened by the overgrown vineyard that fills the vessel's deck and climbs its mast that they jump overboard. Once in the water, the pirates are miraculously transformed into dolphins.[45]

As indicated by this tale of Dionysus, our kinship with the dolphin has been represented in literature and art since ancient times. Because of the dolphin's intimate interaction with people, our fascination with them continues today. There are some researchers in the scientific community who believe dolphins may actually be more intelligent than humans. Traditionally, dolphins are associated with water and signify the power of metamorphosis and regeneration. Dolphins are seen not only as divine emblems of wisdom and prudence, but also as symbols of duality and transformation.[46]

6.44 Dionysus in a boat (Greek drinking cup, 540 B.C.). Note the seven highly stylized dolphins.

Drawing by George J. Haas. (Image source: *History of Art*, by Janson, page 80.)

Another interesting connection to consider between the dolphin, the triangle, the Sun, and with Dionysus, is that his epithet proclaimed he was none other than the only god who was "twice-born" and was called "the child of the double door."[47] This reference to the double door may just be our first clue to the hyperdimensional power found in the twin triangles of that ancient symbol of the Sun, known most commonly as the Star of David. These same connections surface again 35 million miles away on a plain called Cydonia.

Notes

1. Stanley V. McDaniel referred to this anomaly as a "winged ear" on April 25, 1998, while on an Enterprise Mission discussion board entitled "Cydonia in April." His comments are recorded in a discussion post numbered 448.

2. Stanley V. McDaniel, "Peculiarities At 'Main Pyramid.'" The McDaniel Report Newsletter, May, 7, 1998: http://www.mcdanielreport.com. It should be noted that SPSR geologist Harry Moore originally discussed finding this apparently ice-filled crater in a paper presented to the American Geophysical Union (AGU) on May 28, 1998. In his paper Moore reports finding "water ice" in the crater a few weeks before McDaniel. See Stanley V. McDaniel, "SPSR Scientists Present New Cydonia Analysis: Papers Presented at American Geophysical Union Conference." The McDaniel Report Newsletter, 1998: http://www.mcdanielreport.com

3. Mary Ellen Miller, *Maya Art and Architecture* (New York: Thames & Hudson, 1999), 74.

4. Linda Schele and Mary Ellen Miller, *The Blood of Kings: Dynasty and Ritual in Maya Art* (New York: George Braziller, 1986), 50.

5. Ibid., 48.

6. Karl Taube, *The Legendary Past: Aztec and Maya Myths* (Austin: University of Texas Press, 1997), 34.

7. Ibid., 36.

8. Zecharia Sitchin, *The Cosmic Code: Book VI of The Earth Chronicles* (New York: Avon, 1998), 172.

9. Ibid., 171.

10. Linda Schele and Mary Ellen Miller, op. cit., 46.

11. Ingri and Edger Pairn D'aulaire, *The Greek Myths* (Garden City, NY: Doubleday, 1992), 41.

12. Zecharia Sitchin, op. cit., 49.

13. Jeremy Black and Anthony Green, *Gods, Demons and Symbols of Ancient Mesopotamia* (Austin: University of Texas Press, 1995), 27.

14. Ibid., 76.

15. Zecharia Sitchin, op. cit., 49.

16. Jeremy Black and Anthony Green, op. cit., 82.

17. Jean Chevalier and Alain Gheerbrant, *Dictionary of Symbols* (New York: Penguin, 1996), 282.

18. Jeremy Black and Anthony Green, op. cit., 75. The stag refers to the giant fallow deer.

19. Zecharia Sitchin, *The Lost Realms: Book IV of The Earth Chronicles* (New York: Avon, 1990), 183.

20. Mary Miller and Karl Taube, *The Gods and Symbols of Ancient Mexico and the Maya: An Illustrated Dictionary of Mesoamerican Religion* (New York: Thames & Hudson, 1993), 104.

21. Ibid., 184.

22. Linda Schele and Mary Ellen Miller, op. cit., 51.

23. Jonathan Norton Leonard and the Editors of Time-Life Books, *Great Ages of Man, A History of the World's Cultures: Ancient America* (New York: Time Inc., 1967), 30.

24. Douglas Gillette, *The Shaman's Secret: The Lost Resurrection Teachings of the Ancient Maya* (New York: Bantam, 1997), 92.

25. J. E. Cirlot, *A Dictionary of Symbols,* transl. from Spanish by Jack Sage, 2nd ed. (London: Routledge, 1971), 343.

26. Confirmation of these three pictographs was acquired in the 2002 MOC image E02–01847. The presence of a fourth pictograph is evident in the new data: it will be evaluated in a future edition of this book.

27. Douglas Gillette, op. cit., 229. The word *uay* is also spelled *way* and is pronounced "why."

28. Linda Schele and David Freidel, *A Forest of Kings: The Untold Story of The Ancient Maya* (New York: Quill, 1990), 210.

29. Douglas Gillette, op. cit., 39.

30. David Freidel, Linda Schele, and Joy Parker, *Maya Cosmos: Three Thousand Years on the Shaman's Path* (New York: Quill, 1993), 50.

31. Ibid., 130.

32. Jean Chevalier and Alain Gheerbrant, *Dictionary of Symbols* (New York: Penguin, 1996), 1016.

33. The Maya calendar has been translated into our modern calendar. August 13, 3113 B.C. is considered the starting date of the Maya calendar denoted in its western equivalent.

34. Linda Schele and Peter Mathews, *The Code of Kings: The Language of Seven Sacred Maya Temples and Tombs* (New York: Touchstone, 1998), 37 and 217.

35. Ibid., 37.

36. Linda Schele and David Freidel, op. cit., 209.

37. Linda Schele and Peter Mathews, op. cit., 214.

38. Jean Chevalier and Alain Gheerbrant, op. cit., 1019.

39. Douglas Gillette, op. cit., 200.

40. The "Fort" or "Fortress," which is located at the northeast corner of the City complex, was discovered around 1981 by Richard C. Hoagland in the original *Viking* data prepared by DiPietro and Molenaar. Hoagland believed this anomaly to be a structure with three straight walls enclosing a triangular interior space and a series of crisscrossing geometric striations. See Richard C. Hoagland, *The Monuments of Mars,* 4th ed. (Berkeley: North Atlantic Books, 1996), 14–15, and caption from plate 6.

41. Mary Miller and Karl Taube, op. cit., 90.

42. Personal conversation with Joel Skidmore via e-mail at Mesoweb, "Ask the Archaeologist." http://www.mesoweb.com. It is not known if the Olmec culture made any distinction between fish and aquatic mam-

mals; therefore, they may have seen the dolphin as a fish and been intrigued by its ability to survive out of water.

43. Jean Chevalier and Alain Gheerbrant, op. cit., 303.

44. Richard Hoagland states that this coat hanger feature is part of a ruined tetrahedral structure that lies directly under the mesa below the Face, which is located at 19.5 degrees off the "D&M–Face" line. See The Enterprise Mission article, "Massive Tetrahedral Ruin" (4/24/98) at http://www.enterprisemission.com/images/mars/tetra.jpg.

45. Robert Graves, *The Greek Myths* (Wakefield: Moyer Bell, 1994), 106.

46. Jean Chevalier and Alain Gheerbrant, op. cit., 303.

47. Robert Graves, op. cit., 56.

CHAPTER **7**

The Key of Solomon

The Sidonians (Cydonians)

The megalithic structures located throughout the Cydonia complex on Mars are truly enormous and awe-inspiring; many of them encompass acres of land and go on for miles. The Earth has numerous examples of megalithic structures that also confound logic, although they are not nearly as large.

Early in the second century B.C., Antipater of Sidon compiled a list of architectural wonders of his time that became known as the Seven Wonders of the World. His list included the Pyramids of Egypt, the Hanging Gardens of Babylon, the Statue of Zeus at Olympia, the Temple of Diana at Ephesus, the Mausoleum at Halicarnassus, the Colossus of Rhodes, and the Pharos of Alexandria. Of all these magnificent feats of architecture, only the pyramids survive.

Aside from this "best of" list, there are many other examples of buildings and monuments constructed of gigantic blocks of stone, sometimes weighing hundreds of tons, that can still be found scattered throughout the ancient world. Some of these extant structures include the Stone Age temple in Malta, the legendary megalithic Stonehenge complex in England, the splendid mud-brick ziggurats of Mesopotamia, and the extraordinary mountain-size pyramids of Teotihuacan and Sipan in the Americas.

Another of these structures is the incredible megalithic platform called the Temple Mount on Mount Moriah in Jerusalem. The origins of its mysterious builders may offer a clue toward understanding the massive structures at Cydonia. Who had the ability and the knowledge to construct such incredible architectural marvels that have perplexed scholars for centuries? The answer may lie within an ancient fraternity known as the "Master Builders." One source of information on the Master Builders can be found with the mention of the Sidonians of the Bible. It is reported that they built the Temple of Solomon on the Temple Mount, with materials supplied by King Hiram of Tyre. Unfortunately, after the Temple's completion the Sidonians mysteriously faded from the biblical record. Who were these great architects of the past and where did they go? As you will see, our quest for an answer leads us right back to Cydonia.

The word "Sidonian" is based on a generic Greek name for the Phoenicians and/or Canaanites, who were also called the Zidonians in the Bible.[1] They established a vast maritime enterprise handling cedar wood throughout ancient Lebanon. These Zidonians were the inhabitants of the ancient city of Zidon, which was a Phoenician city on the eastern coast of the Mediterranean Sea near Lebanon. The Greeks called this city Sidon, a word that signifies fish or fishing.[2]

If we take a closer look at the word "Cydonia," we find that it is possibly a corruption of the Greek word "Sidon" or "Sidonia." As stated, Sidon was a major port city of Canaan, and according to the Bible (Numbers 13:28, 33), some of the inhabitants of the land of Canaan at the time of Moses included the Anak, who were the descendants of the Nephilim. The word "Nephilim" has been translated by many as giants, but literally it means "those who were cast down" or "those who descended to Earth." The word originates from the Akkadian term "Anunnaki," which refers to the original gods who came down from the heavens and settled Earth.

In this land of Canaan, at a place called Baalbek nestled in the Cedar Mountains of Lebanon, a huge stone platform even larger than the Temple Mount has existed since antiquity. This was the land of the Canaanite Sun god Ba'al. The Bible calls this place Beth-Shemesh (house/abode of Shemesh). Shemesh was the Sun god of the Sumerians, who were known as the Chaldeans in the Bible. The Greeks called this place Heliopolis (city of Helios), again a reference to the Sun god. The stone platform consists of layer upon layer of huge stone blocks, some weighing hundreds of tons.

Particularly amazing is the Trilithon, which forms the middle layer. It consists of three gigantic quarried rectangular stones with an estimated weight of 1,100 tons each.[3] Who could have produced these immense blocks, transported them high into the Cedar Mountains, and raised them precisely in place? This House of the Sun could only have been built by the Master Builders.

In 1958 the International Astronomical Union (IAU), which is in charge of planetary and satellite nomenclature, arbitrarily titled the patch of land on Mars in which the Face and its enormous surrounding structures reside "Cydonia," a word that phonetically sounds just like Sidonia. The entire span of Martian topographical real estate was titled within the guidelines set forth in the late nineteenth century by the Italian astronomer G. V. Schiaparelli, best known for his observations of lines on the surface of Mars that he called "canali." This tradition suggested thematic titles that included all kinds of epithets culled from the extensive mythos of our past. According to the IAU, the Martian piece of geography titled Cydonia is named after "a poetic term for Crete."[4] In ancient Crete, the modern-day city of Chania was called none other than Cydonia (Kydonia).[5]

Our research found the origins of the word Cydonia in the pastoral lands of Archadea, in a city called Tegea. According to local myths, the founder of this city, Tegates, had several sons by Maera, the daughter of

Atlas. One of the children of this blessed union was called Cydon.[6] When the boys eventually grew into men, they immigrated to the island of Crete, where they founded several cities, all of which were named after them. The ancient Cretan city called Cydonia was named after Cydon, the grandson of Atlas, the bearer of the vault of heaven.[7]

Is the name Cydon a word corruption of Sidon, which is an ancient word for fish? Were the Sidonians of the Bible an overseas colony of the Cydonians that resided at Crete? Whatever their origins may be, the association is intriguing. Historians may claim that the island of Crete has little to do with the city of Sidon or the race of people known as the Sidonians, other than the similarities in their names. However, as you will soon see, the two are very much connected and the word "Cydonia" is a major clue to the unraveling of this code.

The Freemasons

Historically, the Sidonians, the Master Builders of biblical times, are seen as the original architects of the ancient mysteries of "sacred geometry." They are the original keepers of that craft, which later became associated with the Freemasons. It is widely believed that Freemasonry's origins are found in the stories of Solomon and King Hiram in the Old Testament. Although the actual time of the birth of Freemasonry is in question, certain aspects of its ceremony and ritual can be traced back to King Nemrod and the building of the Tower of Babel. It was not until the early eighteenth century that a majority of Masons abandoned the legacy of King Nemrod in favor of King Solomon.[8] Two major doctrines of the Freemasons were the secret knowledge of the science of symbolism, and the application of the principles of geometry.

For students of modern-day Masonic craft, masonry and geometry are identical sciences and the terms are synonymous. This code of ancient symbols and geometry, which is embedded in Masonic rituals, now contains only the remnants of the geometrical secrets of a long-lost science. Three of the most common and most significant symbols adopted by the Masonic craft encode the principles of duality.

The first is the Seal of Solomon (or Star of David), which consists of two interlocking triangles, echoing the double-delta or double-door symbolism discussed in the previous chapter and highlighted in the legend of Dionysus. The second is the compass-and-square emblem, which combines the opposition and transmutation of a square and a circle. The third symbol is the twin columns, erected on the porch of Solomon's Temple and known as Jacobin and Boaz. These pillars emulate the persona of the celestial twins derived from the ancient myth of the Gemini, which represent duality and opposition. Symbolically, these two pillars represent the eternal conflict: one pillar creates and the other destroys.[9]

The symbolic code embedded in these Masonic emblems is a repeated message that speaks of a universal duality. This binary system of opposition is found in abundance throughout Cydonia; it speaks of the amazing geometry involved in the construction there, and provides evidence of a common source of sacred geometry.

The Seal of Solomon

According to the Bible, the Sidonians were ruled by a king called Baal and they worshipped a Sun god by the same name.[10] This connection between the Sidonians and the Sun god is even more interesting, considering that the very name Solomon, which is traditionally associated with peace, actually means Sun. The variant meanings are caused by the two different names associated with Solomon.

The Hebrew name for Solomon is Shelomth, which means "the Peaceful one."[11] However, the translation of the common name of Solomon has the root meaning of sun. The "Sol" portion of Solomon is borrowed from the Latin word "sol," which means sun.[12] The word "sol" is a reference to the Roman Sun god Sol, who is identified with the Greek god Helios (the Sun). The second half of the word "Sol-o-mon" is a derivative of the Greek word "mon," which means one.[13] So we are left with a simple definition of the word Solomon: One Sun.

The graphic design of the Seal of Solomon consists of two equilateral triangles, one pointing up and the other pointing down, creating a six-pointed star. This symbol is considered one of the most powerful magical emblems of all time. It combines two alchemical triangles, one for water and one for fire. The triangle pointing up, or ascending, represents fire, while the descending, or "inverted triangle," signifies water. According to alchemical science, when these two elements intertwine, air and earth are created. Therefore it could be said that the Seal of Solomon represents the "Philosopher's Stone, expressing the union of the four elements of the universe."[14]

The Sun's highest temperature emissions and maximum sunspot activity both occur at approximately 19.5 degrees north and south. If you place two equilateral triangles so that each has a vertex at one of the Sun's poles, the base of each triangle crosses the Sun at 19.5 degrees and you end up with Solomon's Seal (Figure 7.1). The Seal of Solomon was both a symbol of the Sun that held a matrix to the sacred knowledge of tetrahedral geometry, and an extension of a hexagram (a star-shaped figure formed by extending the sides of a regular hexagon to meet at six points).

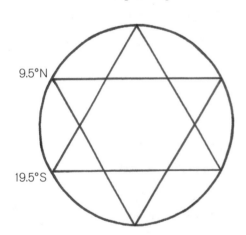

9.5°N

19.5°S

7.1 Two intersecting triangles within a sphere.

Drawing by George J. Haas

Hexagonal Craters on Mars

In examining the Cydonia images taken in April 1998, many anomaly researchers noticed that numerous craters in the Cydonia area seemed to have polygonal shapes. The most notable are hexagonal. On the second Cydonia swath (MOC SP1–23903) taken on April 14, we noticed one of these large hexagonal-shaped structures (Figure 7.2), approximately the size of one and a half football fields, at the south end of the swath. What should have been a round crater was actually a hexagonal walled structure. This is not a shape that comes naturally to a crater. Depending on the angle of impact, a splash pattern should be more or less circular or oval.

A second hexagonal feature is found in a *MGS* image (M22–00378) just below the lozenge-shaped mound, known as Mound P (Figure 7.3). The upper left arm of the structure appears to be bent or partially collapsed. The remaining five straight walls complete a full hexagon.[15] Notice the network of radiating lines that surrounds the six-sided feature. At the bottom of the crater, in the top portion of the hexagon, is an anomalous set of

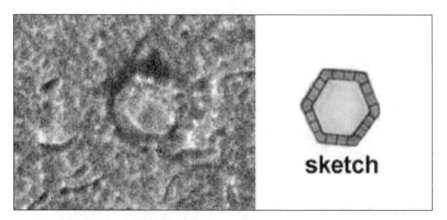

7.2 Hexagonal crater on Mars.

LEFT, Context image of hexagonal crater. Note the hexagonal shape of the walled area of the crater.

RIGHT, Drawing of the hexagonal crater.

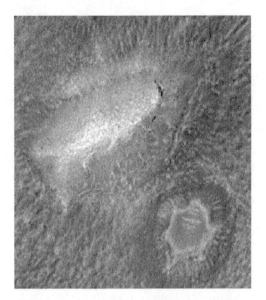

7.3 Hexagonal crater near Mound P (cropped from *MGS* image M22–00378). Note Mound P (upper left), with six-sided hexagonal crater (lower right). Additional confirmation of this hexagonal crater is available in MOC image E0101908.

Enhancement courtesy Keith Laney

parallel lines. These two hexagonal structures confirm, in a symbolic and repetitive way, the messages of Cydonia. A hexagon is a symbol of duality and, according to Carl Jung, the archetypal symbol of opposites.[16]

Solomon's Seal in Mesoamerica

The hexagram also appears as a sacred symbol in pre-Columbian Mesoamerica. The Maya created elaborate costumes and masks as aids to supernatural empowerment. They wore feathered costumes with wicker-framed back racks, decorated with quetzal feathers that identified the wearer as the Principal Bird deity. On a Mayan terra-cotta sculpture depicting a warrior with shield (Figure 7.4), the warrior looks a lot like a feathered angel similar to the winged deities of Mesopotamia (Figure 6.13). The figurine appears to be wearing the costume of the Principle Bird deity and displays a pendant with a hexagonal design.[17]

The internal physics of a hexagon is that of two superimposed, equilateral triangles placed together to form a six-pointed star. In turn, each point adheres to the form of a hexagram (Figure 7.5). The resulting emblem is the Seal of Solomon, a symbol of the sacred geometry that expresses duality and, most important, the Sun.[18]

In the mythology of the Maya, the Principle Bird deity is also seen as the Celestial Bird known as Itzam-Ye, or Seven Macaw. In the Maya *Popol Vuh,* the Itzam-Ye bird poses as a false Sun and is eventually defeated by

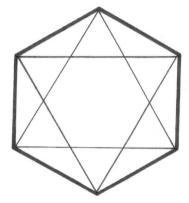

7.4 Maya warrior dressed as the Principle Bird deity (wearing hexagonal pendant).

Drawing by George J. Haas. (Image source: *National Museum of Anthropology, Mexico City*, by Ragghianti and Collobi, pages 107, 108.)

7.5 Hexagram within hexagon. Note that the Seal of Solomon symbol fits within the hexagon.

Drawing by George J. Haas

the Hero Twins.[19] The sculpture of the feathered warrior is not an angel, but a personification of the Itzam-Ye bird possessing the Sun, in the form of a hexagonal pendant.

Additional evidence that the Maya were familiar with hexagonal geometry is demonstrated in the design of what is called the Maya Star Shield, which dates to 1000 B.C. (Figure 7.6A).[20] This six-pointed star emblem, within a ribbon motif, was found at Uxmal, a city in the Yucatan. Its composition incorporates a Seal of Solomon with a double circle in its center. The star is surrounded by another circle with twelve petals and has a feathered tail that is nearly identical to a winged rosette motif found in Mesopotamia (Figure 7.6B).

7.6A Feathered rosette motif. Maya Star Shield with Seal of Solomon and feathered tail (found at Uxmal, in the Yucatan).

Drawing by George J. Haas. (Image source: *Ancient Past of Mexico,* by Reed, page 12.)

7.6B Mesopotamian winged rosette. Note the near-identical tail formation.

Drawing by George J. Haas. (Image source: *Stairway to Heaven,* by Sitchin, page 94.)

7.7A Four Regions emblem. Note the overall Seal of Solomon shape of the Sumerian Four Regions emblem.

Drawing by George J. Haas. (Image source: *Wars of Gods and Men,* by Sitchin, page 180.)

7.7B Seal of Solomon.

Drawing by George J. Haas

Sitchin has suggested that the post-diluvial emblem used by the Sumerians to signify The Supreme Place of the Four Regions, which was located at the heart of modern-day Jerusalem (also known as the Navel of the Earth), may have been a forerunner of the Seal of Solomon (Figure 7.7A).[21] Notice the six points, or rays, of the crown-like Four Regions emblem and how it echoes the design of the Seal of Solomon (Figure 7.7B). The Four Regions were ruled over by the Sumerian Sun god Shamash (Shemesh.)

The Mark

The second most important symbol associated with sacred geometry is found on the Masonic coat of arms, which combines a builder's compass and set square. The simple mechanics of this emblem symbolize the essence of duality: the compass forms a circle, while the set square forms a square (when mirrored). The circle and square are opposites, expressing duality.

On the third Cydonia swath (MOC SP1–25803), some unusual markings were noticed on the surface next to the Main Pyramid, just above the head of the dolphin (Figure 7.8). Applying the mirroring technique, an incredible image was produced resembling the familiar symbol used as a

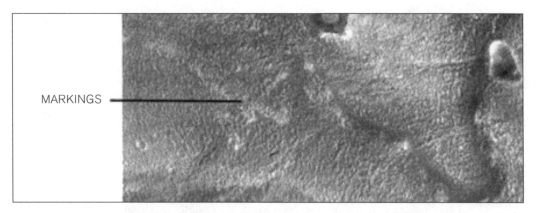

7.8 Context image of the Mark (detail of third Cydonia swath, MOC SP1–25803). Note the elaborate markings on the Martian surface.

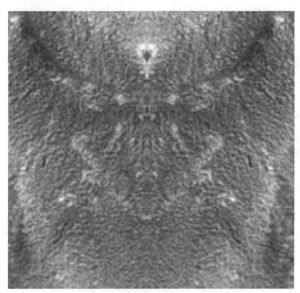

7.9 The Mark (mirrored). Note the rotational swirl effect at the top of the compass.

7.10 Freemasonry symbol of the compass and square.

Drawing by George J. Haas

sacred emblem of the Freemasons—a compass and square! (See Figures 7.9 and 7.10.)

These two forms were also of great importance to the Maya. Among the many accouterments found in the tomb of the great Maya leader King Pacal of Palenque were a jade cube, found in his right hand, and a jade sphere, found in his left hand (Figure 7.11). These objects echo the duality of the square and the compass.

The Maya believed in a dualist cosmology that proclaimed each individual was part of a similar piece of Creation. The psyche of the Maya was a dualist perception of "I am you" and "You are me," which, like their gods, represented the opposing forces of nature.[22]

A Mayan linguist, Domingo Martinez Paredes, maintains that the Maya had a cosmic principle of movement and measurement that symbolized a

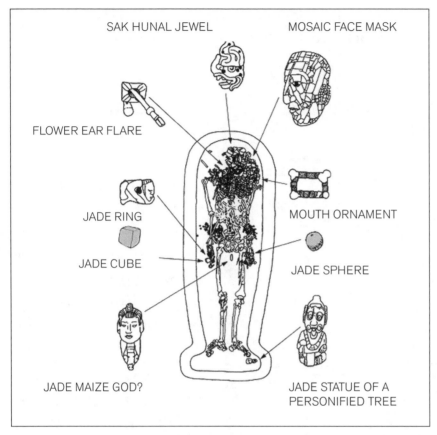

SAK HUNAL JEWEL

MOSAIC FACE MASK

FLOWER EAR FLARE

JADE RING

MOUTH ORNAMENT

JADE CUBE

JADE SPHERE

JADE MAIZE GOD?

JADE STATUE OF A
PERSONIFIED TREE

7.11 Sarcophagus of the Maya leader King Pacal (his remains and accouterments). Note the jade cube in the right hand and the jade sphere in the left hand. (Gray highlights added by the authors.)

Drawing by Linda Schele, © David Schele. (Courtesy of Foundation for the Advancement of Mesoamerican Studies, Inc., www.famsi.org.)

dualist view of the universe that they called Hunab Ku (Sole Measure Giver). They believed in a dynamic dualism in which the whole material world was part of a cosmic mathematical order. The Hunab Ku was symbolized by a square in a circle (Figure 7.13A). According to Martinez, the Hunab Ku and the Masonic compass and square are synonymous concepts that symbolize the standards of the "Great Architect of the universe."[23]

A similar idea in which a sole god creates the entire universe is illustrated by a thirteenth-century French illuminated manuscript of the Bible. On one of the colorful panels, almighty God is cast as the Great Architect of the universe (Figure 7.12). Compass in hand, in the act of creation, God contemplates the circumference of the cosmos.

In the Maya *Popol Vuh,* the formation of a square is described as one of the initial actions performed by the gods in order to create the cosmos. The creator gods arranged the Four Corners of the heavens in the shape of a square, while the Earth formed a circle below it. The Maya, with the aid of a simple cord, adopted these sacred measurements and incorporated them into their daily lives. According to Mayanist Christopher Powell, the square was the fundamental shape found in Maya geometry, and served as the model from which all Creation was generated.[24]

To form the Hunab Ku, the Maya would first form a square and quarter it and then place a cord at the center and stretch the cord along a 45-degree angle to the corner of the square. Then by running the cord completely around the square they were able to form a perfect circle (Fig-

7.12 God as the Great Architect of the Universe (thirteenth-century French illuminated manuscript of the Bible, Pierpont Morgan Library). Note the compass being used to measure the universe.

Drawing by George J. Haas. (Image source: *Mythology,* by David Leeming, 1976, page 152.)

ure 7.13B).[25] This very geometric shape of a circle and a square set up the concept for the duality of the universe. All aspects of their world and humanity were split into paired deities that are complementary, as evident in their myths and bifurcated art work. The Maya also incorporated a similar technique of measurement in their architecture to achieve the sacred proportion known as the "golden mean."[26]

Just like the Seal of Solomon, the origins of the Hunab Ku might also be traced back to Mesopotamia. The opposite version of the Hunab Ku, symbolized by a circle within a square, is recorded on a 3,800-year-old Babylonian tablet featuring mathematical concepts and geometric exercises[27] (Figure 7.13C). Is this another case of cultural diffusion in the Americas from Mesopotamia, or another clue pointing our way all the way back to Cydonia?

The same instructional diagram that was used by the Maya (and the Sumerians) to create a circle with a square can be found on a coin from the Cretan city of Cydonia (Figure 7.14). On the reverse side of this coin is a geometric diagram of a quartered square with a diagonal marker in its lower left quadrant. This diagonal marker signifies the act of drawing a cord from

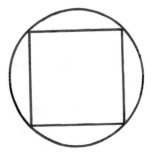

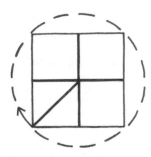

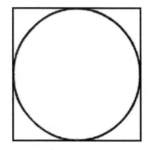

7.13A The Maya Hunab Ku (square within a circle).

Drawings by George J. Haas

7.13B Quartered square (diagram for circling the square).

7.13C Sumerian geometry (circle within a square).

7.14 Cydonia coin (Kydonia, Crete, circa second century B.C.). Note the symbol on the reverse side (on the right) with the diagonal marker within the quartered square.

Drawing by George J. Haas. (Image source: Classical Numismatic Group, by A. J. Gatlin, lot number 362.)

the center of the square to its lower corner and pulling it around the outside of the square, thereby forming a perfectly proportioned circle. The diagram's unique design creates an analogous symbol to a Hunab Ku, where the outer rim acts as a circle. Does this diagram declare that the inhabitants of Cydonia were Master Builders, and provide additional evidence that the Master Builders known as the Sidonians in the Bible were one and the same?

The obverse side of the coin features the head of Apollo, the Sun god who was known as the destroyer in the Iliad.[28] Notice the strong profile and the wing motif incorporated into his hair. Apollo was known for his prophetic abilities and his associations with the arts, music, and mathematics, and his appearance on the face of this coin may be another reference to the art of sacred geometry. In the Iliad, both Apollo and Poseidon are credited by Homer with the building of the walls of Troy.[29] Furthermore, the Greek poet Callimachus described Apollo as a great builder. In one of his poems he proclaims that Apollo "delights in the construction of towns of which he himself lays the foundation."[30]

In this context, where Apollo is portrayed as the Great Architect of Cydonia on a commemorative coin, his attributes are set in opposition: he is seen as both the destroyer and the builder. This said, we must remember Apollo's most common associations, which are his gift for prophecy and his portrayal as a Sun god—Helios. In searching for the origins of the Greek god Apollo, many scholars have made note of his resemblance to the Sumerian Sun god Shamash, who was also a god of prophecy.[31] You may recall that the Sidonians of the Bible worshipped a Sun god called Baal while they were building Solomon's Temple, and it is the word "Solomon" that just happens to finds its root in the Greek word "Helios," and Helios is the god that was also known as Apollo.

Like the simple graphic design of the Hunab Ku, the quartered-square diagram encodes the duality of the circle and square. Both geometric gestures symbolize the Great Architect of the universe. This matrix of duality has become a redundant exercise throughout Cydonia and is expressed in the "cut in half" geoglyphic compass and square that we have identified as the Mark (Figure 7.9). Its presence further establishes a common analog to the duality of opposing forces. If these opposing terrestrial builders known as the Cydonians (of Crete) and the Sidonians (of Lebanon) are indeed related to a common culture, then maybe they are also connected to the Cydonian architects on Mars. If true, the implications are staggering. If not, we are still faced with several questions: With regard to this vast Martian corpus, which is based on a codex of half and split geoglyphs, why are we presented with a "half glyph" of an elaborately designed compass-and-square emblem above the head of a full-bodied profile of a dolphin? And why isn't the dolphin presented as a "cut in half" geoglyph?

Within an eminent structure that borders the dolphin, known as the Fortress, we may find the answer.

Notes

1. William Smith, LLD, *A Dictionary of the Bible* (Nashville: Thomas Nelson, 1997), 766. The Greeks also called the Phoenicians the "phoinikes," which means "red people." See Rick Gore, "Who Were the Phoenicians? New Clues from Ancient Bones and Modern Blood," *National Geographic* 206, no. 4 (October 2004), 34.

2. William Smith, op.cit., 764.

3. Zecharia Sitchin, *The Cosmic Code: Book VI of The Earth Chronicles* (New York: Avon, 1998), 208, 209.

4. For planetary nomenclature, see: http://wwwflag.wr.usgs.gov/USGS-Flag/ Space/GEOMAP/PGM_home.html.

5. http://www.newadvent.org/cathen/04581b.htm. Also see *The Catholic Encyclopedia,* Volume IV Copyright 1908 by Robert Appleton Company, Online Edition Copyright 2003 by Kevin Knight.

6. Pierre Grimal, *The Dictionary of Classical Mythology* (New York: Blackwell Reference, 1986), 270.

7. Ibid.,120. Also see Carlos Parada, *Genealogical Guide to Greek Mythology:* http://homepage.mac.com/cparada/GML/GGGM.html.

8. Daniel Beresniak, *Symbols of Freemasonry* (Singapore: Barnes & Noble Books, 2003), 26.

9. J. E. Cirlot, *A Dictionary of Symbols,* translated from Spanish by Jack Sage, 2nd edition (London: Routledge, 1971), 116.

10. William Smith, op. cit., 766.

11. Ibid., 643.

12. Robert K. Barnhart, *The Barnhart Dictionary of Etymology* (New York: H. W. Wilson, 1988), 1031.

13. The Lex Files, *Basic Greek Elements List A–Z* (Senior Scribe Publications, 2003). The Greek word "mon" means one, alone, single; a number used as a prefix. Online, see http://www.lexfiles.com/basic-grk-m-z.html.

14. Jean Chevalier and Alain Gheerbrant, *Dictionary of Symbols* (New York: Penguin, 1996), 930.

15. Confirmation of this hexagonal crater is available in the 2002 MOC image E0101908.

16. Jean Chevalier and Alain Gheerbrant, op. cit., 504.

17. Carlo Ludovico Ragghianti and Licia Ragghianti Collobi, *Great Museums of the World: National Museum of Anthropology, Mexico City* (New York: Newsweek, Inc. and Arnoldo Mondadori Editore, 1970), 107–108.

18. Jean Chevalier and Alain Gheerbrant, op. cit., 504.

19. David Freidel, Linda Schele, and Joy Parker, *Maya Cosmos: Three Thousand Years on the Shaman's Path* (New York: Quill, 1993), 70–71.

20. Carl G. Liungman, *Dictionary of Symbols* (New York: W. W. Norton & Company, 1994), 301.

21. Zecharia Sitchin, *The Wars of Gods and Men: Book III of The Earth Chronicles* (New York: Avon, 1985), 180.

22. Adrian G. Gilbert and Maurice M. Cotterell, *The Mayan Prophecies: Unlocking the Secrets of a Lost Civilization* (Shaftesbury, Dorset: Element Books, 1995), 78.

23. Peter Tompkins, *Mysteries of the Mexican Pyramids: Dimensional Analysis on Original Drawings by Hugh Herleston, Jr. and Historic Illustrations from Many Sources* (New York: Perennial Library, 1976), 282–283.

24. Linda Schele and Peter Mathews, *The Code of Kings: The Language of Seven Sacred Maya Temples and Tombs* (New York: Touchstone, 1999), 35.

25. Christopher Powell, "The Shapes of Sacred Space," lecture at 19th Annual Maya Weekend, University of Pennsylvania Museum, March 24, 2001. This same technique can be used to produce a circle within a square.

26. Linda Schele and Peter Mathews, op. cit.

27. Editors of Time-Life Books, *Lost Civilizations: Mesopotamia: The Mighty Kings* (Alexandria, VA: Time-Life Books, 1995), 16. A photograph of the tablet including the geometric square and circle is found on page 16.

28. Although the caption placed under the coin identifies the head as a "nymph" (on the website in which the coin is listed), based on comparative examples of Roman coins the head is most likely the head of Apollo. Many Roman coins depict Apollo with a winged motif, as opposed to the Greek god Hermes who is portrayed with a winged helmet.

29. Richmond Lattimore, *The Iliad of Homer* (Chicago: The University of Chicago Press, 1951), 180.

30. Richard Aldrington and Delano Ames, transl., *The Larousse Encyclopedia of Mythology* (New York: Barnes & Noble, 1994), 115, 116.

31. Ibid., 113.

CHAPTER 8

NASA Captures the Fortress

It's a Fortress!

Located in the northeastern portion of the City complex is a geometrical structure that appears to rest on a triangular supportive platform that some have suggested is the remains of a collapsed pyramid. This wedge-shaped enigma was labeled the "Fortress" by Richard Hoagland in the early eighties after he discovered it in a set of SPIT enhancements of *Viking* data provided by DiPietro and Molenaar. This most remarkable structure has become the focus of much debate within the Mars research community, and is considered by Mark Carlotto to be even "more intriguing than the Face"[1] (Figure 8.1).

The inner framework of this fortified structure consists of an almost almond-shaped object that gives an impression of having a tail fin attached to its southernmost side. At the center of this feature is a half-moon-shaped crater that appears to be truncated, forming an eye. This soft feature is contrasted by three almost-straight walls that connect in a zigzag pattern along the adjacent right side. A dark, triangular enclosure or "courtyard" is formed within the structure, suggesting the interior space of a walled compound (Figure 8.2).

On the top ledge of one of the elevated walls along the northeastern side of the structure's perimeter is a feature that Carlotto has described as

8.1 Context image for the *Viking* Fortress. Note the Fortress (boxed in white) is located at the most northeastern portion of the City Center at Cydonia.

Viking orbiter image courtesy of NASA

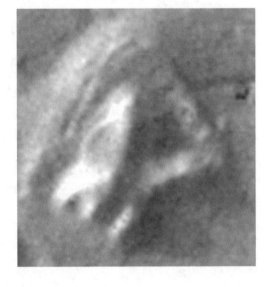

8.2 The *Viking* Fortress (detail of *Viking* 35A72). Note the appearance of straight wall-like features and a dark triangular enclosure.

Viking orbiter image courtesy of NASA

"regularly spaced markings or indentations."[2] These features were so intriguing and geologically perplexing that he appealed to NASA for a higher-resolution image before he could determine the cause of such patterns and subtle details.

A small portion of the Fortress was eventually captured by the MOC camera in the third Cydonia swath taken in late April 1998. Although the southern end of the Fortress is barely noticeable, at the top of the swath the image does reveal a large pictograph of a dolphin resting right next to it (Figure 6.40). The outline of the dolphin pictograph is so close to the Fortress that it may actually be an intricate part of the same structure. For us, the actual significance of this pictographic dolphin would remain a mystery until further pictures were taken of the entire Fortress. Unfortunately, two more years would pass before the public was made aware that NASA had obtained additional photographs of this highly enigmatic structure.

It's Not a Fortress!

On April 5, 2000, NASA released eight new high-resolution (narrow-angle) MOC images of the Cydonia region that had been taken over the previous two years. Included in this unprecedented release were two MOC swaths that contained two additional views of the Fortress (M09–05394 and M04–01903). Each of these new swaths captured only a portion of the Fortress, so to compensate for the missing data, the individual swaths had to be spliced together to make a full image of the structure.[3] Figure 8.3 is a composite image prepared by Mark Carlotto.

Initially, the new images of the Fortress were met with immense disappointment and disillusionment among most supporters of the intelligent-design hypothesis. Although the new high-resolution pictures of the structure were still full of anomalous features, the so-called "blueprint," which included the courtyard and walled boundaries, was no longer visible.

Critics now felt that Carlotto's "regularly spaced markings" turned out to be only the remains of eroded craters, while Hoagland's "courtyard" was found to be only part of the structure's elevated platform. The only wall to remain visible was the straight edge running along the base of the eastern side of the structure's main platform. Although the mesa no longer resembled a fortress, Carlotto stated that the mesa still retained its anomalous character, with a crisper view of new detail.

In the aftermath—just as they had moved on after the release of the second Cydonia swath, it wasn't long before most researchers set aside the Fortress as yet another misreading of inadequate *Viking* data. We, on the other hand, had a completely different take on this new *MGS* image of the Fortress.

One of the most unusual aspects in this new image was the carved trench or pathway radiating from the eastern side of the Fortress. This odd crack

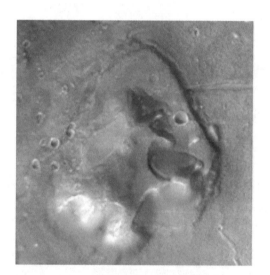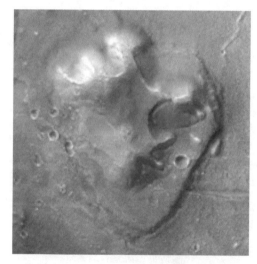

8.3 MOC image of the Fortress (composite image of M09–05394 and M04–01903). LEFT, Original orientation.

RIGHT, Inverted. Note the road or pathway leading into the Fortress along the straight-edged wall on the right side.

Composite image courtesy of Dr. Mark Carlotto

in the terrain appears to run in a straight line from the darkened edge of Car-lotto's main wall right out toward the Face (Figure 8.3B). If this feature is a fault line or crack in the surface, there is no evidence of any residual distur-bance to the wall of the Fortress. Was this trench a road or possibly the remains of a transportation system leading into an underground complex? Right from the start, we believed there was much more here than first meets the eye!

It's Another Split Face!

In examining this new high-resolution image of the Fortress closely, we found the clarity of its internal features to be most revealing. The first thing we did was to rotate the image 180 degrees and back again, in an attempt to find any recognizable features that might lead to an understanding of this structure's amazing complexity. After inverting the image, we noticed what appeared to be a large "fang" protruding from part of an upper lip of a snarling mouth (Figure 8.4).

Adjacent to the fang were the rem-nants of a broad, human-shaped "nostril" sitting below a large, closed "eyelid." Above the fang and to the left we saw a long, smooth, wing-shaped "eye" that occupied the same area where the almond-shaped fea-ture was seen previously in the *Viking* image. It quickly became apparent that this wasn't the remains of a ruined and abandoned Fortress; this was another two-faced geoglyphic structure.

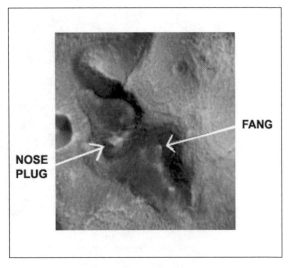

8.4 The fang, with snarling lip (detail of the MOC image of Fortress). Note the large fang, the partial nostril formation, and nose plug.

With the aid of an enhancement, by image specialist Keith Laney,[4] we mirrored the left side of this compelling structure. What emerged was the portrait of a now familiar snarling jaguar in its anthropomorphic aspect (Figure 8.5).

Notice the pronounced feline muzzle and the snarling aspect of the mouth. The fangs and the mitten-shaped "tear band"[5] become even more

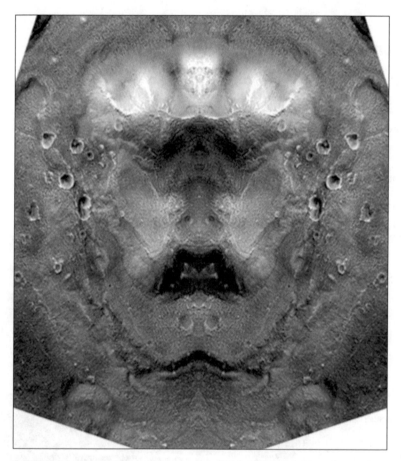

8.5 The Fortress mask (snarling jaguar, mirrored left side). Note the squinting eyes with flame-shaped eyebrows, the nose, the fangs and snarling aspect of the mouth complete with turned-down corners. A winged bat mask occupies the crested headdress area.

Enhancement courtesy of Keith Laney

pronounced in Laney's enhancement. There also appears to be a mask occupying the crested crown area of the forehead. Two small "ears," complete with "pelt marks," formed by a few small, scattered crater marks, flank the highly reflective crown at the top of the head. The shape of the ear resembles the ear found on the head glyph of the anthropomorphic Bearded Jaguar god mentioned in Chapter 1 (Figure 1.22).

The flaming eyebrows are formed by what appears to be a composite glyph of feathered serpents that sit atop the long wing or mitten-shaped tear band. A flat pug nose is seen above the snarling lips, which are noticeably turned down at the corners. The two protruding fangs that are visible in the partially opened mouth emphasize the snarling aspect of an Olmec were-jaguar. A chin or beard completes the shape of the face, which includes a crater forming a chin ornament (Figure 8.6).

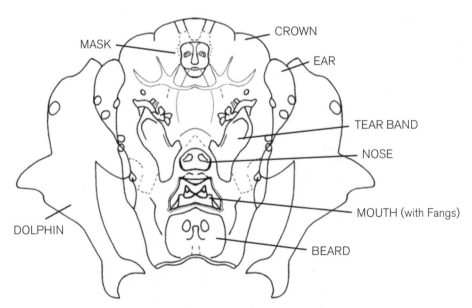

8.6 Fortress mask with dolphins (snarling jaguar, mirrored left side). Note the fangs, the mitten-shaped tear band, and the flaming eyebrows, and how the dolphin pictograph frames the head of the snarling jaguar.

Analytical drawing by George J. Haas

The large pictograph of the dolphin, which was discussed in Chapter 6, frames the head of the snarling jaguar like an immense headdress. The dolphin alongside the snarling jaguar mask resembles the Olmec headdress that featured the arching dolphin seen in the previous chapter and shown again in Figure 8.7. Notice how an E-shaped tear band covers the eye and cheek of the Olmec face on the hacha; compare it to the wing or mitten-shaped tear band in the Snarling Jaguar mask at Cydonia.

The dolphin-adorned wreathed headdress also bears a striking resemblance to one of the "changing-face" masks of the Northwest Coast Indians, most notably the work of the Kwakiutl tribe of Vancouver, British Columbia. This terrestrial companion mask was found in the collection of the Museum of American Indians in New York and dates to circa 1850 (Figure 8.8).

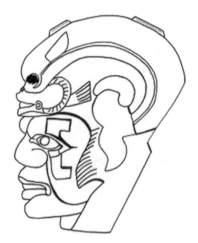

8.7 Dolphin iconography in Olmec headdress (Central Veracruz, Classic period). Votive hacha (stone axe trophy head) with a stylized dolphin headdress.

Drawing by George J. Haas. (Image source: *Mexican National Museum of Anthropology*, by Bernal, page 127.)

8.8 Kwakiutl mask (man's head surrounded by two fish, from British Columbia, circa 1850). Note the open-hand motif on the cheeks and how the two fish frame the head, just as the dolphins do the Martian mask.

Drawing by George J. Haas. (Image source: *Indian Art in North America: Arts and Crafts*, by Dockstader, Figure 91.)

The mask features an elaborately painted human head with two "mythical fish"[6] surrounding its face, in the same manner as the dolphin pictograph framing the mask at Cydonia. This Kwakiutl mask would explain why the dolphin pictograph at the Fortress is presented in a full-body profile as opposed to a half or split image. The dolphin at Cydonia acts as an adornment of the overall Snarling Jaguar mask, just as the fish do with the Northwest Coast Indian mask.

On closer examination of the Indian mask, we notice that it also features an "open hand" motif painted on the cheeks that echoes the mitten-shaped feature seen on the Snarling Jaguar mask on Mars (Figure 8.6). The similarities between this terrestrial Indian mask and the Martian mask are very compelling.

Beyond the dolphin headdress, the facial expression of the Snarling

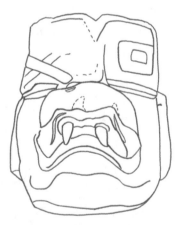

8.9 Olmec were-jaguar with sprout (incised jade head from Las Tuxlas). Note the crown, flame eyebrows, and snarling were-jaguar aspect of the mouth.

Drawing by George J. Haas. (Image source: *Olmec World: Ritual and Rulership*, by Coe, page 106.)

8.10 Olmec head of were-jaguar (La Venta). Note the snarling mouth and closed eyes.

Drawing by George J. Haas. (Image source: *Olmec Art of Ancient Mexico*, by Benson and Fuente, page 176.)

Jaguar mask on Mars shares many design qualities with an Olmec were-jaguar portrait carved on a jade head found at Las Tuxlas, Mexico (Figure 8.9). Notice the crown, the elaborate "flaming" eye treatment, and most notably the snarling aspect of the lips.

Other features include two fangs and a toothless upper gum, with flaring lips turned down at the corners.[7] This snarling aspect can also be found in an Olmec portrayal of a were-jaguar found at La Venta, Mexico (Figure 8.10). Notice how the "pug" nose rests on the upper snarling lip, forming a down-turned mouth, exposing its raw gums and two large fangs.

The Sleeping God

When the right side of the Fortress is mirrored along the same central axis, a second face forms. This companion face to the Snarling Jaguar depicts a triangular mask that we called the "Sleeping God." This almost heart-shaped face is composed of a set of two large, closed eyelids, a broad nose with flaring nostrils, a cylindrical nose ornament, ribbed "ear bars," and a fan-shaped crescent worn as a headdress (Figure 8.11). An analytical drawing is provided for clarification of these unique facial features (Figure 8.12).

Xipe Totec

The Sleeping God mask at Cydonia resembles the Aztec god Xipe Totec. The gold buckle in Figure 8.13 features the head of Xipe Totec, called "Our Lord of the Flayed One." The headdress, the large, closed, almond-shaped eyes, the shape of the nose along with the crest, the narrow chin, and the overall shape of the face are remarkably similar to the Sleeping God mask. Another (most notable) similarity is the nose plug found on both!

In the Aztec culture, Xipe Totec was the god of Spring and also seen as the "redeemer of suffering." It has been noted that his face was pockmarked

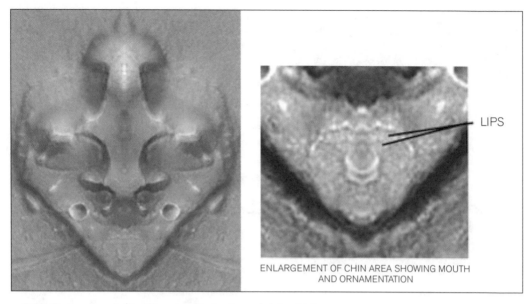

LIPS

ENLARGEMENT OF CHIN AREA SHOWING MOUTH
AND ORNAMENTATION

8.11 Fortress mask (the Sleeping God, mirrored right side).

LEFT, Full image. Note the large eyelids, broad nose complete with nose ornament
and nose plugs, V-shaped chin, and fan-crested headdress.

RIGHT, Detail of mouth and chin. Note the bat-wing-shaped mouth and the
V-shaped chin.

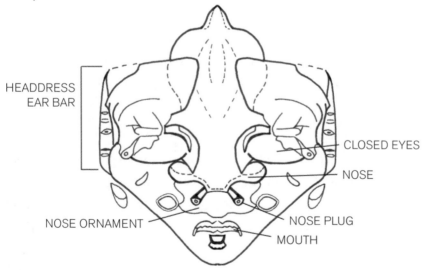

HEADDRESS
EAR BAR

CLOSED EYES

NOSE

NOSE ORNAMENT

NOSE PLUG

MOUTH

8.12 Fortress mask (the Sleeping God, mirrored right side).

Analytical drawing by George J. Haas

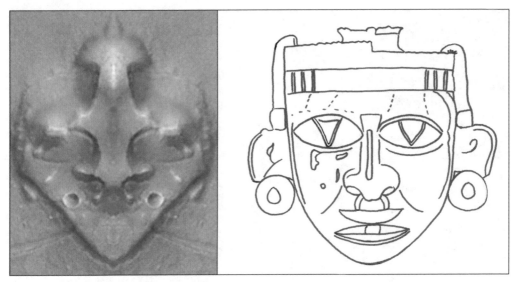

8.13 Sleeping God comparison (Earth and Mars).

LEFT, The Martian Sleeping God (right side of the Fortress mask, mirrored).

RIGHT, Xipe Totec (gold buckle). Note the headdress, closed, almond-shaped eyes, the shape of the nose along with nose plug, the crest, and narrow chin on both images.

Drawing by George J. Haas. (Image source: *Aztecs of Mexico: Origins, Rise and Fall of the Aztec Nation,* by Vaillant, plate 46.)

and pustule-ridden, giving him a hideous appearance. During the springtime he flayed his own skin "in order that the active growing principle hidden within matter could be freed."[8] The flaying of the skin suggests a ritual of agricultural renewal.

In Aztec ceremonies, the human skins of captives were worn by warriors until they rotted off, revealing the renewed man within. This act mimics the evolution of the corn seed, as it breaks out of the Earth's crust and emerges as a new tender shoot.[9] In examining closely the Sleeping God side of the Fortress mask, we noticed the crater acting as a "pockmark" below the nose and various "tear drop" markings throughout the face. These may be the remnants of the suffering and pustule-ridden features that were so intensely characterized by Xipe Totec.[10]

The Aztec cult of Xipe finds its origins in the Olmec and Zapotec cultures. Artifacts have been found that depict priests or shamans covered by the flayed skins of animals and humans. Among Zapotec Indian artifacts, the most dominant type of cloak or cape worn by shamans is the flayed pelt of the jaguar.[11] The Olmec influence of pelt coverings that feature snarling were-jaguars can be seen in this fine example of a clay figurine of a sleeping priest wearing a feline pelt (Figure 8.14). This one Olmec figurine (among other examples) unites both aspects of these early ideas of the flayed god and the were-jaguar.

8.14 Olmec priest. Note the snarling were-jaguar aspect of the figurine's face, the closed eyes, and the flayed feline pelt on his back.

Drawing by George J. Haas. (Image source: *Mexican National Museum of Anthropology*, by Bernal, page 40.)

The companion portrait to the Snarling Jaguar mask on Mars is the Sleeping God, whom we now believe represents the Martian equivalent of the Aztec god called Xipe Totec, Our Lord of the Flayed One. The snarling Olmec figurine of the priest wearing the caped jaguar pelt is a pre-Aztec representation of the flayed god Xipe as a were-jaguar.

With this in mind, we feel confident in proclaiming that the Fortress mask at Cydonia represents a two-faced geoglyph of the same two attributes seen in Mesoamerican iconography of the Snarling Jaguar and the Flayed One. Again, not only is another two-faced geoglyphic structure present within the Cydonia complex; we have found that each half of this bifurcated face is intimately and iconographically connected.

Notes

1. Mark Carlotto, *The Martian Enigmas, A Closer Look* (Berkeley: North Atlantic Books, 1997), 31.

2. Ibid.

3. To fully appreciate the structural composition of the Fortress, we have to first create a composite image. This is done by taking a portion (one half) of the Fortress from *MGS* image M09–05394 and connecting the other half, at the precise points, from *MGS* image M04–01903. As a result of fusing each half of the Fortress, a representation of the entire structure is created. Questions on how the decision was made to photograph this highly anomalous structure in two separate halves can only be addressed by NASA.

4. Keith Laney (a.k.a. Bullitt) is a digital-imaging and software applications specialist and MOC image processor for the NASA–Ames' MOC MER2003 Landing Sites Project and has provided digital enhancements for NASA's online publication of *Apollo Over the Moon: A View from Orbit.* He is currently working on the complete Apollo program image archives for NASA/JPL.

5. A tear band is a decorative element placed on the face of many Mesoamerican and Peruvian masks. Tear bands can take on many motifs, such as a rectangular bar or a zigzag pattern representing the tracks of tears.

6. Frederick J. Dockstader, *Indian Art in America: The Arts and Crafts of the North American Indian* (Greenwich, Conn.: New York Graphic Society, 1961), caption for Figure 91.

7. Michael D. Coe, *The Olmec World: Ritual and Rulership* (Princeton: Princeton University Press, 1996), 120.

8. Irene Nicholson, *Mythology of the Americas* (London: Hamlyn, 1970), 224.

9. Ibid.

10. This mask of the flayed god represents a face after having been skinned, so the V shape of the chin may be the result of sagging facial tissue, caused by the absence of skeletal support. A missing jawbone would also result in the drooping of the skin. This characteristic is most often seen in shrunken heads.

11. George C. Vaillant, *Aztecs of Mexico: Origins, Rise and Fall of the Aztec Nation* (Garden City, NY: Doubleday, 1950), 614.

CHAPTER 9

The Enlightened One: Cosmic Dragons and Vision Serpents

The Parting of the Second Cydonia Swath

After the swirling sands of excitement ("Cydonia mania") generated from the third *MGS* swath had settled down, we had time to relax enough to remove our blinders and refocus. We began what would become a slow and methodical review of many of our earlier discoveries, which needed to be recorded and catalogued. We also began to re-examine the second swath (MOC SP1–23903) that had been so quickly set aside by almost everyone involved in anomaly hunting, including ourselves, when the third swath was released.

Most of the researchers seemed to be looking for the devil in the details on these Cydonia images. However, we decided to step back and examine the overall swath. Having learned that the original Face on Mars was not the only megalithic structure consisting of partial images to be found in Cydonia, we took an educated look at the second swath from a new perspective. One of the most intriguing oddities in the second swath was a line of contrast running vertically and, curiously, right down the center of the entire image. This demarcation line appeared to be an intentionally etched par-

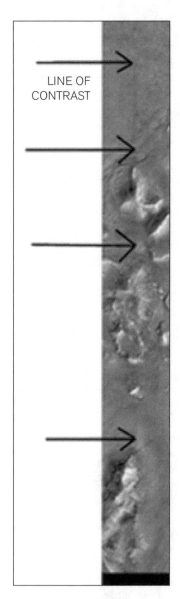

9.1 Context image. Second Cydonia swath (MOC SP1–23903, 1998). The horizontal black band is data that was lost in transmission from Mars to Earth.

Courtesy of NASA/JPL/Caltech

tition that traversed the diverse topography of this Martian landscape, separating the second swath into two sides (Figure 9.1).

Not knowing for sure if this line was just an image artifact of the Mars Orbital Camera or an actual intended marker, we performed a mirror split on each side of this partition line to see what, if anything, was produced. As soon as we examined the mirrored swath, the images began to emerge. We were now looking at the second swath in a new and most revealing light. The entire swath contained a multitude of split pictographic images that aligned themselves one after another down the partition line. The swath read like an immense totem and, just as we had witnessed with the original Face, some of these pictographs were also inverted. Once again, we were confronted with what appeared to be sacred pages from the lost book of the Maya that were "hidden from the searcher and the thinker."

It took many months (and many sleepless nights) to interpret the pieces of this enigmatic puzzle and fit them together. However, once the structures began to reveal their hidden secrets, their significance began to fall into place. Because of the pictographic form that these immense geoglyphic images display and their complex, interactive relationship, we are presenting them here not in the order in which they were found, but in a manner that hopefully will be the easiest to follow.

The Fetus

At the southern portion of the swath lies a mega-lithic structure of which only a portion was captured by the camera on board the *Mars Global Surveyor*. The partition line runs along the edge of this structure, indicated by the black arrow in Figure 9.2. The black band at the bottom is data NASA reported lost in transmission.

When the west side of the image was mirrored along the partition, a recognizable image was not immediately apparent. It wasn't until the section was inverted or turned south to north that an image of a baby materialized. It appeared we were seeing a gestational offering of an entire human being. The most notable feature of this infant is that it demonstrates the large head-to-body ratio typical in the normal development of a human fetus (Figure 9.3). The head features a nose, eyes, and a swollen upper lip, which overhangs the mouth. There is also evidence of swelling under its slanted eyes. The child holds a ring in its large hands against its chest and has its knees pulled up in a fetal position. The legs extend downward, ending with two little feet. The overall figure also seems to have a transparent quality to its skin that is most evident in its head. The lower portion of the fetus is framed by a jagged fan- or vulva-shaped membrane, which gives the appearance that a child is actually being born. Figure 9.3

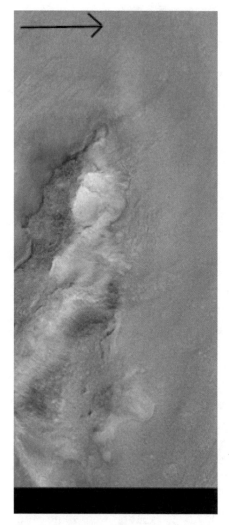

9.2 Context image. Southern portion of MOC SP1–23903 (from Figure 9.1). The black arrow points out the line of contrast. The black band at the bottom is data NASA reported lost in transmission.

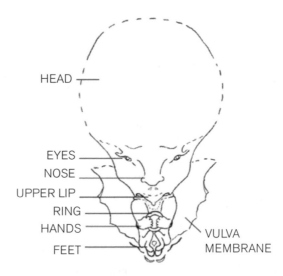

HEAD

EYES
NOSE
UPPER LIP
RING
HANDS
FEET

VULVA
MEMBRANE

9.3 Martian fetus.
LEFT, Mirrored image of fetus on Mars.
RIGHT, Analytical drawing of Mars fetus. Note the slanted eyes, the swelling under the eyes, and the representation of transparent skin in the face and skull.
Analytical drawing by George J. Haas

shows a composite image of this complete fetus, along with an analytical drawing for reference.

In the former Olmec city of La Venta (in the present-day state of Tabasco, Mexico), on the south end of the city's approximately north-south axis, archaeologists found three large sculptures of Olmec fetuses. These sculptures were some of the "thousands of crafted, semiprecious stone objects deposited in spatially significant arrays below the surface of the ceremonial courtyards."[1] Olmec researcher Carolyn Tate and her colleague, Gordon Bendersky, suspected that more than two dozen Olmec sculptures previously thought to be dwarfs, dancers, or depictions of the supernatural were-jaguar were actually figurines of human fetuses. Eleven specialists, including obstetricians, gynecologists, neonatologists, perinatologists, and embryologists, confirmed their speculations. These specialists determined that the sculptures were so startlingly precise that they were able to identify the gestational age of the fetus represented.[2] Some of the unique fetal characteristics demonstrated in these Olmec figurines are the large head, with a head-to-body ratio of about 1:3 or 1:4, slanted eyes, and swelling under the eyes.[3]

Another feature remarkably represented in some of the sculptures is "cutaneous pellucidity." This is found in fetuses between 12 and 25 weeks and refers to the transparency of the skin that allows the veins underneath to be seen.[4] Most intriguing is that many of the fetuses have adornments, including pierced ears, ear spools, and incised iconography, and carry sacks containing maize seeds, indicating that the figures hold a religious significance.

Some of the sculptures display iconography of pelt and head tufts, characteristics of the Olmec Dragon of Creation, while others wear helmets

9.4A Olmec fetus, standing figure with incisions (Chiapas provenance, 900 B.C.). Note the extended lip.

Drawing by George J. Haas. (Image source: *Olmec World:Ritual and Rulership,* by Coe, page 220.)

9.4B Olmec fetus. Dragon incisions (found on the fetus).

Drawing by George J. Haas. (Image source: *Olmec World: Ritual and Rulership,* by Coe, page 221.)

worn by players of the ritual Mesoamerican ball game, symbolizing death and rebirth.[5] Figures 9.4A and 9.4B show an Olmec sculpture of a fetus with the incised iconography of a dragon on its head and down its back.

After studying the apparent symbolic nature of these Olmec sculptures, Carolyn Tate posed the questions: "As new souls, were fetuses seen as messengers from the Otherworld? Or did a fetus effigy serve as a powerful symbol of the physical and spiritual transformative capacity of the human being?"[6] According to what we have found on Mars, the answer is yes to both questions.

Buddha in the Lotus Bloom

The transparency of the fetal head at Cydonia becomes even more important when one focuses on the small figure in the center of the skull. This figure, seated in an almost meditative posture, has an uncanny resemblance to a Buddha (Figure 9.5). In particular, this figure resembles the Buddha Maitreya or Future Buddha. The Martian Buddha appears to be sitting on a lotus throne within the open petals of a larger lotus bloom. He sits in a typical lotus position and even the curled fingers on his hand can be seen. An analytical drawing of this structure appears in Figure 9.6.

Generally, the lotus bloom, which has many botanical varieties and has evolved with many spiritual associations, is a symbol of water, fertility, and creation. The first lotus is said to have arisen from the primordial waters at the beginning of time, and is linked with the fertility of the gods. In a Hindu myth, Brahma comes forth from a lotus blossom growing from an umbilical cord that connects to the navel of Vishnu, who sleeps on a vast ocean of cosmic water. In Maya creation myths, the Maize Tree, which is the embodiment of First Father, emerges from a water-lily blossom. Its roots are compared to an umbilical cord that is connected to a celestial body of water, the Milky Way.[7]

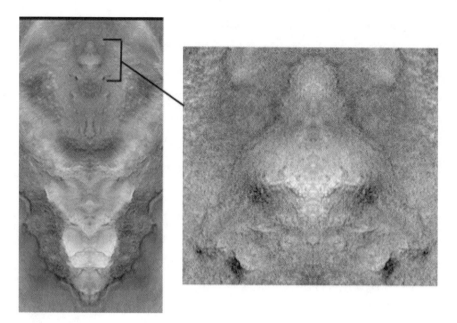

9.5 Mars Buddha.

LEFT, Mirrored fetus image from Figure 9.3 (showing location of the Buddha-like figure).

RIGHT, Buddha-like figure (detail). Note the meditative posture.

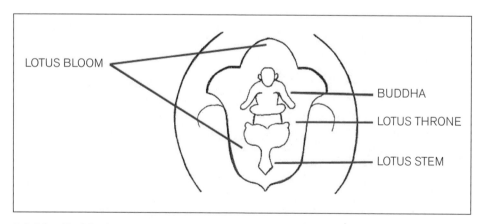

9.6 Mars Buddha in a lotus bloom.

Analytical drawing of Figure 9.5 by George J. Haas

Throughout Asian cultures, the Buddha is typically found sitting on a lotus throne, as seen in the Japanese sculpture in Figure 9.7. Traditionally, the Buddha sits in the middle of an eight-petaled lotus, perched at the hub of an eight-spoked wheel, symbolizing the eight points of the compass. Because the stalk of the lotus is seen as a phallic symbol on which this great wheel rotates, the lotus is seen as the central point of the world axis.[8] In a sense, the lotus acts as the Maya World Tree, which also marks the point of the world axis.

The lotus, which is the most important meditative flower throughout the Middle and Far East, can also be equated with the white water lily of Mesoamerica.[9] Many Maya sculptures have been discovered with a similar god-within-a-flower motif, just as seen in eastern cultures (Figure 9.8). Notice the meditative, Buddha-like presence of this Maya figure within the lily blossom. According to Maya mythology, the god Kakoch created the

9.7 Buddha (Amida of the west). Note the aura of the Buddhahood and the lotus throne.

Drawing by George J. Haas. (Image source: *History of Far Eastern Art*, by Lee, page 305.)

9.8 Maya Buddha in a white water-lily bloom.

Drawing by George J. Haas. (Image source: *Mysteries of the Ancient Americas*, by Editors of Reader's Digest, page 26.)

first water-lily plant that floated on the primordial waters and it was from its blossom that all the other gods were born.[10]

In the Maya culture, a water-lily motif frequently occurs as a design element around pottery and buildings. The image in Figure 9.9 is from a wall relief found at Palenque. In this decorative band, a god is being born as its head rises up out of the lily.

This analogy of the water lily as a vehicle of birth is fostered by its resemblance to a vascular organ that develops within the human uterus. The size and shape of a water lily and its long stem resemble the umbilical cord attached to the placenta and connected to the fetus in the womb.[11] If this is an accurate interpretation of the water lily/lotus analogy, then the embryonic fluid could be seen as the primordial waters that sustain this sacred and anointed flower.

In ancient Egypt, where the lotus was the symbol of Upper Egypt, the

9.9 Maya head rising from lily blossom (detail).

Drawing by George J. Haas. (Image source: *The Blood of Kings,* by Schele and Miller, 1986, Figure 24, page 46.)

9.10 Egyptian lotus (a god's head being resurrected from a lotus blossom).

Drawing by George. J. Haas. (Image source: *Dictionary of Symbols,* by Biedermann, page 213.)

hawk-headed god Horus is sometimes depicted seated on a lotus bloom, which is seen as a symbol of rebirth.[12] In the Egyptian Book of the Dead, a similar idea to the Buddha in a lotus bloom is represented as a god's head rising out of a lotus blossom, here symbolizing resurrection (Figure 9.10).[13]

The meditative, mirrored image of the Martian Buddha in Figure 9.5 not only resembles the Buddha Maitreya; it also bears a striking likeness to a Mayan sculpture found at the ancient Maya site of Quirigua in Guatemala. Called Altar P, this odd, saucer-shaped sculpture carries the image of a seated figure. According to J. Eric S. Thompson in his book *The Rise and Fall of Maya Civilization,* this large, carved river stone features a Buddha-like figure seated in the open mouth of a "celestial dragon" (Figure 9.11). The Buddha's relationship with this dragon motif will prove to be more than just a coincidence.

Across the world, far from the ancient ruins of the Maya, in the flood plains of the Mekong River in Cambodia, lies the ancient temple city of Angkor Wat. Housed in one of the temples is a somber statue that features another aspect of the Buddha relating to this dragon/serpent motif. In this image, a Buddha sits on a coiled serpent while the snake's head rises up behind him, exposing its swelling hood (Figure 9.12). This sculpture depicts a famous episode in the Buddha's life when he falls victim to a storm that rages with torrential rains for a whole week, while he sits meditating. The serpent king, Nagas Muchilinda, comes up through the Earth to cover the Buddha as he meditates in the rain. The serpent coils his body around and around to form a seat for the Buddha, then swells his hood to provide shelter from the streaming rain.[14]

The coiled serpent acts as the lotus throne, while the expanded hood acts like a lotus bloom protecting the Buddha from the elements of the world. Just as the Maya figure sits in the mouth of the Celestial Dragon at

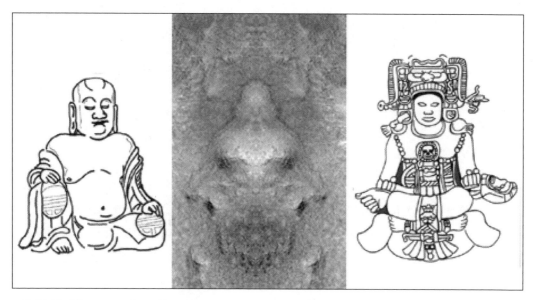

9.11 Buddha comparison. Note the amazing similarities between all three images.

LEFT, The Buddha Maitreya.

Drawing by George J. Haas

CENTER, Mars Buddha (mirrored image).

RIGHT, Maya Buddha (Altar P, Quirigua). Note the apron that hangs down in front.

Drawing by George J. Haas. (Image source: *Rise and Fall of the Maya Civilization,* by Thompson, Plate 32.)

9.12 Buddha sitting on Muchilinda (Cambodia, twelfth century A.D.). Note the serpent hood.

Drawing by George J. Haas. (Image source: *Angkor: The Hidden Glories,* by Freeman and Warner, page 128.)

Quirigua (Figure 9.11) and remains unharmed, the Angkor Wat Buddha is also at one with his own version of the Celestial Dragon.

Researcher Graham Hancock maintains that the temples at Angkor Wat were laid out in a pattern reflecting the serpent-shaped constellation of Draco.[15] The function of this transfiguration between monuments and stars is thought to be a celestial blueprint for uniting the heavens with Earth.

A similar relationship with Draco is found at an Olmec site in Takalik Abaj, Guatemala. Archaeologist Christa Schieber de Lavarreda discovered that by tracing the alignment of standing stones found behind a platform called Structure 7 with a fragmented stela of a Celestial Serpent, the monuments pointed directly toward the heart of the serpent-shaped constellation of Draco.[16] The real significance of this alignment with Draco, in both Asia and Mesoamerica, lies in its relationship with its companion star Polaris, the axis mundi.

The Double One

Carolyn Tate had described some of the Olmec fetus sculptures mentioned above as being adorned with the inscribed iconography of pelts and head tufts, identifying the markings with the Olmec dragon. We saw an example of this in Figure 9.4. According to Tate, this Celestial Serpent was created by the Olmec "to give form to the powers inherent in the sky and Earth." As a cosmic symbol, the Olmec dragon allowed shamans to open pathways between the natural and supernatural world.[17]

When we take another look at the mirrored fetus image on Mars, we notice that the body of the fetus transforms into the head of the Celestial Dragon (Figure 9.13). The dual aspect of this geoglyph is revealed by a technique mentioned in Chapter 3 called contour rivalry, which allows the viewer to perceive two intentional readings within one image. This formu-

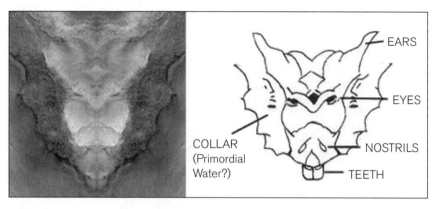

9.13 Mars dragon.
LEFT, Mirrored image of the dragon.
RIGHT, Analytical drawing of dragon.
Analytical drawing by George J. Haas

lated dragon head appears to come complete with eyes, nostrils, ears, teeth, and a feature that may be a collar, or perhaps represents the dragon's wake as it floats in the primordial waters.

In the *Popol Vuh*, the Maya refer to a hideous, crocodile-like dragon called the Zip Monster (Zipacna), which existed long before the first humans were created. He existed at the time when the Earth was still forming. The Zip Monster was the son of the Itzam-Ye bird and was known as the "Maker of the Mountains."[18] According to Mercedes de la Garza, the Maya dragon "symbolizes a supreme sacred energy permeating the entire cosmos ..." and was also related to the Sun, blood, semen, and corn.[19]

On the side of a Late Classic Maya vessel is a profiled representation of one of these reptilian dragons (Figure 9.14). Notice the long muzzle, protruding teeth, and collar. The ancient Maya creation date of 3113 B.C. is also recorded on this vessel.

A similar Olmec sculpture from 900 B.C. found in Mexico is described as a figure riding a crocodile (Figure 9.15). It looks very much like a small child or fetus on the back of a Zip Monster or dragon.

9.14 Maya dragon: the Zip Monster. Note the long muzzle, protruding tooth, and collar.

Drawing by George J. Haas. (Image source: *Legendary Past: Aztec and Maya Myths,* by Taube, pages 73, 74.)

9.15 Figure riding a crocodilian creature (Mexico, 900 B.C., carved serpentine).

Drawing by George J. Haas. (Image source: *Olmec World: Ritual and Rulership,* by Coe, page 185.)

According to Chevalier and Gheerbrant, the dragon "is a celestial symbol, of the life force and power of manifestation, ejaculating the primeval waters or the World Egg."[20] They also contend that the blossom of the lotus flower is equated with the hatching of this World Egg; the lotus is seen as the first pure life form that arises upon the vastness of the primeval waters: "... the tight bud is the precise equivalent of that egg, and the hatching of the egg corresponds to the opening of the bud. Both are the realization of the potential contained in the first seed."[21] The common symbolism of dragon and lotus appears to have a relationship that stretches back to the far reaches of time. Francis Huxley in his book *The Dragon* describes the dragon as follows:

Dragons come in every size and, since they are notoriously promiscuous, in a variety of surprising shapes. The reason for this must be looked for in their parentage, in the elemental mother and father of them all who is responsible for the beginning of things—when all is still in flux and the elements have not yet distinguished themselves.... Now it is obvious that when nothing is as yet created, all that a fiery eye can see in the abyss is its own reflection. The sight of this is said to inflame the dragon with the envious desire to engulf it, which it does both by coupling with it sexually and by devouring it whole. For this reason the First Dragon is held to be a One made of two genders—though some say it is made of four androgynous couples—who lust after each other with such incestuous voracity that they can only maintain their separate natures by changing identity. They do this time after time, and through this process of natural selection the myriad of forms of life arise from the Double One, continuously and without intermission. Since all these forms descend from the Lord of Progeny, they all inherit the fate of that lord, which is to know life, be known by death, and have progeny on their own account.[22]

Ouroboros (the One, the All)

Another representation of this primordial dragon is expressed in the Ouroboros serpent, which forms a circle by biting its own tail. This ancient concept is almost global in its distribution and can be found throughout Europe, Asia, Africa, and the Americas. Its power and appeal can be traced to Masonic iconography and the magical symbols employed in alchemy. The Ouroboros is surely complex in its symbolism, but it is generally seen as representing the cyclic nature of time. In an ancient Greco-Byzantine manuscript known as the Codex

9.16 The Ouroboros (from the eleventh-century Codex Marcianus). Note in the center of the serpent's ring, the words "the One, the All" in Greek.

Drawing by George J. Haas. (Image source: *Magic Symbols*, by Goodman, page 6.)

Marcianus, which dates from the eleventh century A.D., the Ouroboros is illustrated as a serpent that is half light and half dark (Figure 9.16). The dark and light segments of the body allude to time and the opposing principles of night and day, similar to the Chinese concept of yin and yang.[23] The

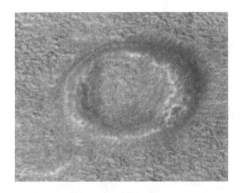

9.17 Mars Ouroboros ("the serpent who swallows his own tail") (from MOC image M10–03053).

Courtesy of Keith Laney

LEFT, Ouroboros discovered on Mars.

RIGHT, Analytical drawing of the Mars image.

Drawing by George J. Haas

notion of time is represented here as a serpent devouring itself, codifying the idea that with every end there is a new beginning. In this instance it is seen as "the One, the All."[24] Carl Jung believes this alchemical legend expressed in the idea of "the One, the All" refers to "Time and Again," and remarks that "The alchemists reiterate that the opus proceeds from the one and leads back to the one."[25] Jung further describes the Ouroboros as the "... dragon that devours, fertilizes, begets, slays, and brings itself to life again. Being hermaphroditic, it is compounded of opposites and is at the same time their uniting symbol."[26]

This whole concept begins to sound a lot like the Olmec idea of the Celestial Serpent.

On one of the later swaths of the Cydonia area released in April 2000 (MOC M10–03053), there appears to be a Martian image of Ouroboros. On the rim of a crater is the distinct image of a serpent swallowing its tail (Figure 9.17). Notice the serpentine body, the head, the mouth, the eye, and the tail. This Ouroboros-shaped crater is found in an area just to the northeast of the Face on Mars.

The Magnificent Twin

The concept that the dragon or serpent is fundamentally connected with DNA, our Creation, and human sexuality is further demonstrated in Zecharia Sitchin's book *The 12th Planet.* Sitchin contends that the original Biblical couple known as Adam and Eve were actually sterile hybrids with no knowledge of sex or gender. It wasn't until they were given the forbidden fruit of carnal knowledge by the serpent in the Garden of Eden that they were able to become human and able to propagate as the gods.[27] The answer lies in our DNA.

Robert Temple points out in *The Sirius Mystery* that the Egyptian gods Osiris and Isis are at times depicted as entwined serpents, as are the founders of Chinese civilization, Fuxi and Cang Jing. On a bas-relief from the second century A.D., Fuxi is paired with his serpentine wife Na Gau and infant, a result of their divine coupling (Figure 9.18). Much like the image of God as the Great Architect of the Universe seen in Chapter 7 (Figure 7.12), Fuxi and Nu Gau are often depicted holding a mason's square and compass.[28] They were said to have repaired the broken heavens and created civilization after the great flood.

This idea of entwined serpent couples also existed in Mesoamerica. A fine example is found in the Codex Mendoza, which features an Aztec couple entwined, not by their tails but by their wedding garments, in an act of "tying the knot" (Figure 9.19). This serpentine ceremony of tying the knot may actually be linked to the mingling of the couple's DNA at conception.

9.18 Entwined serpents Fuxi and Cang Jing (founders of Chinese civilization). The image is from a bas-relief in the Han Dynasty, Wu Liang tomb, second century A.D. Note the mason's square in the right hand of Fuxi.

Drawing by George J. Haas. (Image source: *The Sirius Mystery,* by Temple, page 294.)

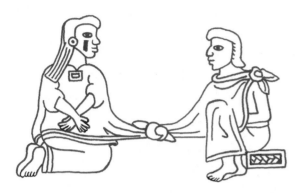

9.19 Aztec couple in the act of "tying the knot" (Codex Mendoza).

Drawing by George J. Haas. (Image source: *Aztecs: Reign of Blood and Splendor,* by Time-Life, page 144.)

9.20 Entwined serpents from Mesoamerica (Classic Maya bowl).

Drawing by George J. Haas. (Image source: *Mythology of the Americas,* by Burland, Nicholson, and Osborne, page 152.)

More traditional images of entwined serpents were also depicted throughout Mesoamerica, as seen in Figure 9.20.

The image of entwined serpents actually dates to the first known civilization, the ancient Sumerians, and is found in the form of the caduceus (Figure 9.21). On the libation cup of the Sumerian king Gudea, the Serpent god Ningishzidda is seen in his dual aspect, represented as a pair of serpents entwined about an axial rod.[29] The caduceus is found in many forms and in many cultures, but its basic principle is two serpents entwined around a central axis. This image was eventually adopted by the Greeks and Romans, in the form of the caduceus of Hermes and Mercury, respectively, as a guiding rod for souls in a quest for rebirth and eternal life.[30] The caduceus has often been used to represent the knowledge of healing, and is still in use today to represent the field of medicine.

In his fascinating book *The Cosmic Serpent: DNA and the Origins of Knowledge,* Jeremy Narby describes the connection between the twin serpents of ancient mythology (Figure 9.22) and the double-ribboned "molecule of

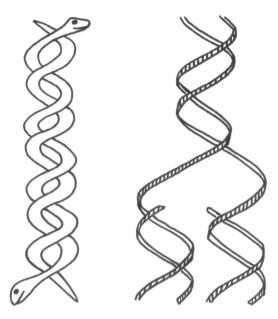

9.21 Early Sumerian version of the caduceus on libation cup (Sumer, 2000 B.C., detail). Note the entwined serpents revealing the dual aspect of the serpent-god Ningishzidda.

Drawing by George J. Haas. (Image source: *The Mythic Image,* by Campbell, page 283.)

9.22 Entwined DNA (serpents).

LEFT, Double helix (represented as a pair of snakes).

Drawing by George J. Haas. (Image source: *Wisdom of the Genes,* by Willis, page 37.)

RIGHT, DNA double-helix ribbon (duplication). Note how the single ribbon breaks into two.

Drawing by George J. Haas. (Image source: *The Double Helix,* by Watson, page 165.)

life," DNA (Figure 9.22). One of the most telling images that supports this connection comes from the Cosmic Serpent of the ancient Egyptians, known as the provider of attributes (Figure 9.23).

The signs above the double serpent mean one ($\mathbf{|}$); several ($\mathbf{|||}$); spirit, double, vital force ($\mathbf{\downarrow}$); place (L); "wick of twisted flax" ($\mathbf{\S}$); and water ($\mathsf{\wedge\!\wedge\!\wedge}$). Under the chin of the second serpent is the ankh, the Egyptian cross meaning key of life. Narby states:

The connections with DNA are obvious and work on all levels: DNA is indeed shaped like a long, single and double serpent, or a wick of twisted flax; it is a double vital force that develops from one to several; its place is water.[31]

Narby quotes molecular biologist Christopher Wills:

The two chains of DNA resemble two snakes coiled around each other in some elaborate courtship ritual.[32]

DNA is a single chain of two interwoven ribbons of nucleotides that are parallel to one another and arranged in a spiral. The nucleotides are connected by four bases, which combine in such a way that the double helix has been compared to a spiraling ladder. Due to the atomic makeup of the bases, they can only be joined in specific pairs. This means one ribbon is the back-to-front duplicate, or mirror image, of the other.[33]

DNA is the informational molecule of life and its very essence consists of being both single and double—like the mythical serpents.[34]

Mischievous twins who can transform themselves into any living creature DNA is the same for all life forms, whether bacteria or human, insect or flower; the only thing that changes from one species to the other is the order of the four bases. It has been estimated that these bases may be joined in over 10,000 billion, billion different combinations in DNA molecules.[35] With this in mind, let's refer back to Huxley's description of the First Dragon earlier in this chapter:

… the First Dragon is held to be a One made of two genders—though

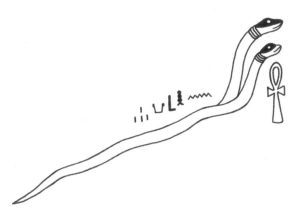

9.23 The Cosmic Serpent of the ancient Egyptians (provider of attributes).

Drawing by George J. Haas. (Image source: after R. T. R. Clark, *The Cosmic Serpent*, by Narby, page 80.)

some say it is made of four androgynous couples (the four bases)—who lust after each other with such incestuous voracity that they can only maintain their separate natures by changing identity. They do this time after time, and through this process of natural selection the myriad of forms of life arise from the "Double One," continuously and without intermission.[36]

With a colorful description settling fresh in our minds of an asexual dragon, seen as a celestial manifestation of the ancient primordial originator of offspring and descendants, our next discovery was not too surprising.

Human Copulation

Within the same mountainous structure that produced the composite Fetus, Buddha, and Celestial Dragon geoglyphs, we also found a composite image with a most intimate display of human sexual coupling (Figures 9.25 and 9.26). As with a page from the Kamasutra, we discovered an enormous pair of human genitals, with the male penis, complete with testicles, apparently penetrating the female vulva. Figure 9.24 shows the location of this amazing image.

As we have repeatedly seen with the Martian geoglyphs, a common archetypal model of duality has emerged. In this case we have the opposite sexes, male and female. The parallel that becomes most apparent in viewing this image is the connection between the mythology of the asexual dragon and human copulation. This thought quickly transcends toward an analogy of human reproduction where the Lord of Progeny

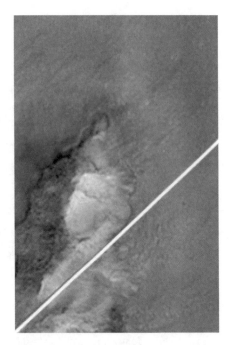

9.24 Context image. Location of split for Figures 9.29 and 9.30.

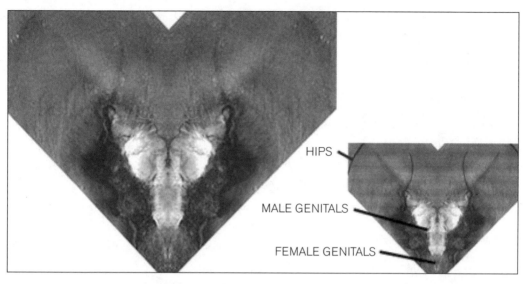

HIPS

MALE GENITALS

FEMALE GENITALS

9.25 Mirrored Mars image of male pelvic area with male and female genitals (highlighted image on the right). See Figure 9.30.

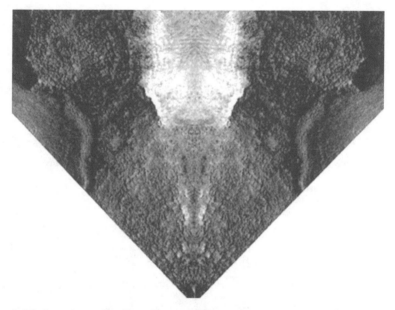

9.26 Sexual coupling (from Figure 9.25, detail). Huxley points out: "Since all these forms descend from the Lord of Progeny, they all inherit the fate of that lord which is to know life, be known by death, and have progeny on their own account."

is similar to a single-celled blastula floating in the primordial waters, unrecognizable as male or female. It is presented as the first divine asexual being, a dragon, or perhaps more fittingly, the Double One.

Biologically, humans start out as asexual beings. In our early development we are only a single cell. After fertilization, our cells, like the androgynous couples of the Lord of Progeny, divide and become two cells. They then split and become four, and so on and so on. This process of cellular division, called mitosis, also happens with an "incestuous voracity."

The distinction of a male or female human fetus does not exist until much later. It is at this point that we undergo sexual separation and are then born. It seems the idea of the dragon, as presented by Huxley, can be seen as a metaphor for DNA, human gestation, and the forbidden fruit of carnal knowledge. It's the dragon or serpent, the giver of that knowledge, that enables us to be separate sexual beings.

Turning our attention back to the geoglyph on Mars, we are confronted with a set of male and female genitals, joined in the act of "cosmic unity," unifying our sexual separation. This act of cosmic unity that occurs between human partners during sexual intercourse, where a sense of oneness is experienced and the individual genders are joined, reunites our androgynous self. This union creates a state of ultimate bliss in which we have a spiritual transcendence of dualities such as male and female, energy and consciousness.[37]

This oneness is expressed in Tantric Buddhism, where the consciousness of enlightenment can be attained through the symbolic union of the two beings called Wisdom and Means. This union is illustrated by the sexual copulation between the male and female organs. In the Tantric painting in Figure 9.27, Bodhisattva Samantabhadra and his partner are posed in a passionate embrace of cosmic unity while sitting on a lotus blossom.[38]

In Chinese erotic literature, the stem of the lotus is seen as a phallic

9.27 The Bodhisattva Samantabhadra (copulating with his partner). Note the aura hood behind him, and the lotus blossom on which they sit.

Drawing by George J. Haas. (Image source: *Mythology, an Illustrated Encyclopedia*, by Cavendish, page 57.)

symbol (representing the penis), while the open petals of the same flower are seen as an archetype of the vulva.[39] In Tantric Buddhism there is an erotic prayer that translates as "Om, jewel in the lotus, amen." This is a Freudian metaphor for the sexual union of the female genitals (the blossom of the lotus, which is the vulva) and the male genitals (the stem, which is the energy).[40] This act of sexual coupling is seen as a re-enactment of the Elemental Mother and Father who are, as Huxley tells us, compelled to unify as one and propagate.

Bloodletting (to Summon the Serpent)

In the Maya culture, when a couple was to be joined in marriage, a bloodletting ritual would take place. This ritual was performed to express pity on the couple by the gods and to summon the ancestors and gods into attendance.[41] Bloodletting by piercing the tongue, genitals, and other parts of the body was an important ritual in Maya culture; it was also a means of summoning the Vision Serpent. It was believed that through the ecstatic rapture achieved during the bloodletting rituals, the occupants of the Otherworld could be summoned.[42]

Droplets of blood were collected in offering bowls filled with sheets of torn paper that were set afire. The Vision Serpent's writhing body would appear, surging upward in the smoke, revealing a supernatural being in its gaping mouth (Figure 9.28).[43]

The Maya were well aware that the loss of large amounts of blood would induce hallucinogenic experiences (caused by the release of endorphins) that would bring them into direct contact with their gods and ancestors.[44] The act of bloodletting was performed by Mayan kings as well as both male and female citizens. The females would pierce their tongue and pull a barbed rope completely through it, while the males would pierce the penis with a lancet made from a stingray spine or obsidian stone.[45]

In the creation story of the *Popol Vuh,* the gods use maize[46] as a form of flesh, mixing it with water for blood, to create humans. In an effort to compensate the gods for their creation, the ritual of bloodletting emanated. Bloodletting was seen as mankind's way of providing sustenance for the gods.[47] Because the first humans were intelligent creatures who could name all the gods, praise them, and, most important, provide them with sustenance (human blood), the gods gave them access to the Otherworld's realm through bloodletting.[48]

9.28 Vision Serpent. Note the god (ancestor) emerging from the jaws of the serpent.

Drawing by George J. Haas. (Image source: *A Forest of Kings,* by Schele and Freidel, Figure 2:3, page 69.)

The Maya had glyphs for both the penis (Figure 9.29) and for blood[49] (Figure 9.30). Notice how the example of a "blood scroll" in Figure 9.30 resembles the vulva feature in the Cydonia structure depicting sexual coupling. This "lazy S"-shaped scroll of the blood glyph resembles the external fold of the vulva seen on Mars. This feature could have a double reading of blood scroll and vulva.

We know that there is sometimes a release of blood by female virgins during intercourse, caused by the breaking of the hymen. The scrolled feature around the tip of the penis could be either the penetrating point of the

9.29 Maya glyph of penis.

Drawing by George J. Haas. (Image source: *Blood of Kings*, by Schele and Miller, Figure B5, page 327.)

9.30 Maya glyph of blood scroll.

Drawing by George J. Haas. (Image source: *Blood of Kings*, by Schele and Miller, Plate 73b, page 94.)

vulva or the blood scrolls from cutting the flattened foreskin with a lancet. This double meaning of sexual coupling and bloodletting is intriguing, considering the explicit imagery of human copulation depicted on Mars.

When the Maya summoned the Vision Serpent through the bloodletting ritual, it was believed that the occupants of the Otherworld, whether dead ancestors or gods, could be summoned and would materialize in ritual objects, mountains, or even within the person performing the ritual.[50] We have seen that within the head of the fetus, which was a ritual object to the Olmec, a small Buddha-like figure in an almost meditative posture is present. Is this enlightened being summoned from the Otherworld by the serpent (Figure 9.5)?

According to Carolyn Tate, "The largest fetal sculptures in Olmec art are the ones found on Structure D-7 at the south end of La Venta's approximately north-south [aligned] axis."[51] At the north end of the complex are three colossal Olmec heads, like the ones we have seen in previous chapters. Archaeologist Rebecca Gonzalez Lauck, in charge of the current excavation of La Venta, states, "In both cases, the sculptures mark the principal area of the city, perhaps signaling the main entrances."[52] She also suggests that with the placement of the fetal sculptures in the south and the colossal heads in the north, "a narrative of growth and transformation of kings from fetus to ancestor" is created.[53]

Is it possible that the same symbolism can be found in the Cydonia complex on Mars? You have just witnessed the unveiling of human pro-creation in these highly stylized fetus/serpent sculptures of the Olmec. Matching geoglyphs have been found along a major axis at the southern end of the second Cydonia swath (MOC SP1–23903). As you will soon witness, when we move northward along this axis, the matrix of our human spirit will be poetically exposed.

Notes

1. Carolyn Tate, PhD, and Gordon Bendersky, MD, "Olmec Sculptures of the Human Fetus," *PARI Newsletter,* 30 (Winter1999). http://www.mesoweb.com/pari/publications/news_archive/30/olmec_sculpture.html.

2. Ibid.

3. Ibid.

4. Ibid.

5. Ibid.

6. Ibid.

7. David Freidel, Linda Schele, and Joy Parker, *Maya Cosmos: Three Thousand Years on the Shaman's Path* (New York: Quill, 1993), 425.

8. Jean Chevalier and Alain Gheerbrant, *Dictionary of Symbols* (New York: Penguin, 1996), 616–617.

9. Hans Biedermann, *Dictionary of Symbolism* (New York: Facts on File, 1992), 212.

10. Mary Miller and Karl Taube, *Gods and Symbols of Ancient Mexico and the*

Maya: An Illustrated Dictionary of Mesoamerican Religion (New York: Thames & Hudson, 1993), 184.

11. David Freidel, Linda Schele, and Joy Parker, op. cit.

12. Anthony S. Mercatante, *Who's Who in Egyptian Mythology,* 2nd ed. (New York: Barnes & Noble, 1995), 88.

13. Hans Biedermann, op. cit.

14. Michael Freeman and Roger Warner, *Angkor: The Hidden Glories* (Boston: Houghton Mifflin, 1990), 128.

15. Graham Hamcock and Santha Faiia, *Heaven's Mirror: Quest for the Lost Civilization* (New York: Crown, 1998), 169.

16. Cliff Tarpy, "Unearthing a King from the Dawn of the Maya Place of the Standing Stones," *National Geographic* 205, no.5 (May 2004), 72–73.

17. Michael D. Coe, *The Olmec World: Ritual and Rulership* (Princeton: Princeton University Press, 1996), 33.

18. Dennis Tedlock, *Popol Vuh: The Maya Book of the Dawn of Life* (New York: Touchstone, 1986), 36.

19. Peter Schmidt, Mercedes de la Garza, and Enrique Nalda, eds., *Maya* (Mexico: Bompiani, 1998), 235–236.

20. Jean Chevalier and Alain Gheerbrant, op. cit., 308.

21. Ibid., 617.

22. Francis Huxley, *The Dragon* (London: Thames & Hudson, 1979), 6.

23. Jean Chevalier and Alain Gheerbrant, op. cit., 728.

24. J. E. Cirlot, *A Dictionary of Symbols* (New York: Barnes & Noble, 1995), 246.

25. Carl Jung, *Psychology and Alchemy* (Princeton: Princeton University Press, 1980), 293.

26. Ibid., 372.

27. Zecharia Sitchin, *The 12th Planet: Book I of The Earth Chronicles* (New York: Avon, 1976), 368–372.

28. Robert Temple, *The Sirius Mystery: New Scientific Evidence of Alien Contact 5000 Years Ago,* 2nd ed. (Rochester: Destiny, 1998), 290.

29. Joseph Campbell, *The Mythic Image* (New York: MJF, 1974), 283.

30. Ibid.

31. Jeremy Narby, *The Cosmic Serpent: DNA and the Origins of Knowledge* (New York: Tarcher/Putnam, 1999), 102.

32. Ibid., 92.

33. James H. Otto, Albert Towle, and Trueman J. Moon, *Modern Biology* (New York: Holt, Rinehart, Winston, 1963), 610.

34. Jeremy Narby, op. cit., 90.

35. James H. Otto, Albert Towle, and Trueman J. Moon, op. cit.

36. Francis Huxley, op. cit.

37. Ajit Mookerjee, *Kundalini: The Arousal of Inner Energy* (New York: Destiny, 1982), 64.

38. Richard Cavendish, *Mythology: An Illustrated Encyclopedia* (New York: Barnes & Noble, 1993), 55–57.

39. Jean Chevalier and Alain Gheerbrant, op. cit., 616.

40. Hans Biedermann, op. cit., 213.

41. Linda Schele and Mary Ellen Miller, *The Blood of Kings: Dynasty and Ritual in Maya Art* (New York: George Braziller, 1985), 176.

42. Linda Schele and David Freidel, *A Forest of Kings: The Untold Story of the Ancient Maya* (New York: Quill, 1990), 70.

43. Linda Schele and Mary Ellen Miller, op. cit., 182.

44. Ibid., 177.

45. Ibid., 180.

46. Since maize is a plant that cannot seed itself, the cutting off of the head of the maize (by removing the husk) is necessary if the seeds are to be removed and planted for the next crop. To the Maya this action symbolizes the decapitation of the Maize god and the perforation of the penis. The story becomes an analog of sustenance, sacrifice, and rebirth all in one. See Linda Schele and Mary Ellen Miller, op. cit., 184.

47. Linda Schele and Mary Ellen Miller, op. cit., 176.

48. Ibid.

49. Although the Maya glyph for penis adheres to a standard anatomical form, the glyph for blood has many varying shapes and styles of flowing loops and curls.

50. Linda Schele and Mary Ellen Miller, op. cit., 325–326.

51. Carolyn Tate, PhD, and Gordon Bendersky, MD, op. cit.

52. Ibid.

53. Ibid.

CHAPTER 10

The Mountain of Life, Death, and Resurrection

The mountain of Mars (or Janus) which rises up as a mandorla of Gemini is the locale of Inversion—the mountain of death and resurrection; the mandorla is another sign of Inversion and of interlinking, for it is formed by the intersection of the circle of the Earth with the circle of heaven. This mountain has two peaks, and every symbol or sign alluding to this "situation of Inversion" is marked by duality or by twin heads.

—Marius Schneider

The Partition

According to the Maya *Popol Vuh,* at the beginning of the creation of the world the heavens were divided into eight directional parts, while the Earth was established with its original four corners.[1] These standard points were set forth at the exact time that the World Tree, which was located at the center of the universe, was erected by First Father. This great single tree, located at the axis mundi, or center of the world, was called the Wacah Chan or "raised-up-sky."[2]

On the trunk of this cosmic tree was placed a "tzuk" glyph, which consisted of a half face of "God C." This tzuk head has a mirror glyph in his

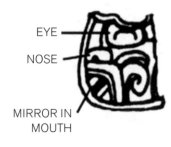

EYE

NOSE

MIRROR IN
MOUTH

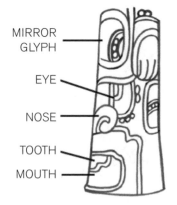

MIRROR
GLYPH

EYE

NOSE

TOOTH

MOUTH

10.1 Maya tzuk glyphs.
LEFT, Found in the belly of the
Itzam-Ye bird in Figure 10.2.
RIGHT, Found on the World Tree
in Figure 10.11.

Drawing by George J. Haas. (Image
source: *A Forest of Kings*, by Schele
and Freidel, 1990, Figure 2:1, p.67.)

forehead that marked the point of the original partition and separation of the Earth and sky.[3] Figure 10.1 shows two variant tzuk heads. It appears that not only did the Maya incorporate half faces in their artwork, they also provided a mirror glyph that prescribed access.

At the top of the World Tree sat the famous Itzam-Ye bird. He was the mischievous bird who proclaimed himself the Sun and was later killed by the Hero Twins to prepare for the Fourth Creation. The Itzam-Ye bird, like the World Tree, was also a signifier of the divisions of the world.[4] A tzuk sign can also be found in the abdomen of this bird, where it, like so many Maya glyphs, has a variant meaning. In this case the tzuk sign is to be read as "belly."[5] In this context, the belly signifies the navel or center of the world from which the four points of the compass radiate.

On a monument found in Kaminaljuyu, Mexico, there exists an extraordinary carving of one of these great magical Itzam-Ye birds that most efficiently demonstrates this variant meaning of the "partition" glyph (Figure 10.2). Among the representations of this avian creature, this one is unique. The Itzam-Ye bird stands here almost camouflaged among a hodgepodge of decorative motifs and glyphs that appear to have no real order. However, upon closer examination, its hidden composite arrangement begins to unfold.

The most obvious bird features are the two wings that spread out from the central figure. A partition or belly sign can be seen in the lower left side of the drawing, identifying this great bird as an Itzam-Ye bird. When the entire image of the Itzam-Ye bird is partitioned by mirroring each side of the bird

10.2 Carving of Itzam-Ye bird (Kaminaljuyu, Mexico). Note the complex interweaving in the design.

Drawings by Linda Schele, © David Schele. (Courtesy of Foundation for the Advancement of Mesoamerican Studies, Inc., www.famsi.org.)

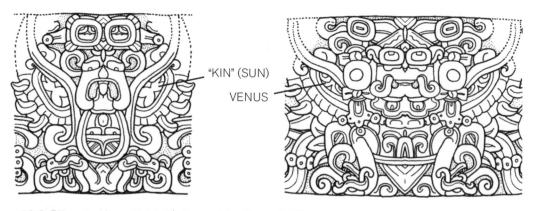

"KIN" (SUN)

VENUS

10.3 Bifurcated Itzam-Ye bird (mirrored from Figure 10.2).
LEFT, Mirrored left side (with kin or Sun glyph).
RIGHT, Mirrored right side (with Venus glyph).

along the central axis of the tzuk sign, two completely different birds emerge (Figure 10.3). The left side of the carving features an open-mouthed bird, whose wings expose a "kin" glyph signifying his earlier disguise as the Sun; a writhing double eagle-headed serpent is at its feet. On the right side, the Itzam-Ye bird transforms into a tentacle-footed bird with a pair of long, curl-

231

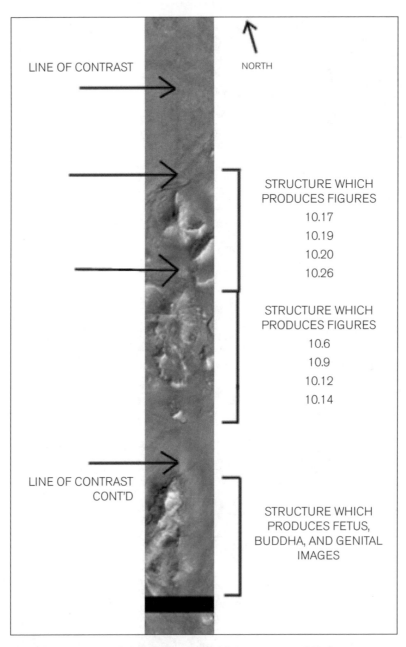

10.4 Context image. Original *Mars Global Surveyor* second Cydonia swath (MOC SP1–23903), 1998. The black line is data that was lost in transmission from Mars to Earth.

Courtesy of NASA/JPL/Caltech

ing tusks projecting out of his mouth. These tusk features may actually represent the "breath of life" sign, seen on the mouth of the Sun god. Notice the goggled eyes, the elaborate ear flares, and the Venus glyph on his chest. A second Venus glyph is found hidden in his wing.

First Lord and the Quadripartite Monster

As we suggested at the end of the last chapter, we believe the partition that runs down the center of the second Cydonia swath for its entire length may actually be the equivalent of the central axis found at La Venta. We believe this also signifies the partition at the center of the world, which First Father set forth at the beginning of creation. For when we move northward on this partition line to the next mountain, more stunning images emerge out of this unlocked codex.

When the eastern side of the swath in Figure 10.4 is mirrored and the image is inverted, looking from north to south, the resulting image amazingly seems to match the description of the Maya god known as First Lord, wearing a Quadripartite Monster headdress (Figure 10.6). The Maya god First Lord (GI) is described by Schele and Freidel as human in aspect and distinguished from his brothers by a shell ear flare, a square eye, and a fish fin on his cheek. He "... [o]ften wears the Quadripartite Monster as his headdress and is associated with the Waterbird."[6] Linda Schele's beautiful drawing of the Maya First Lord wearing a Quadripartite headdress (Figure 10.5) illustrates its complexities.

10.5 Maya First Lord (wearing Quadripartite Monster headdress).

Drawing by Linda Schele, © David Schele. (Courtesy of Foundation for the Advancement of Mesoamerican Studies, Inc., 2002.)

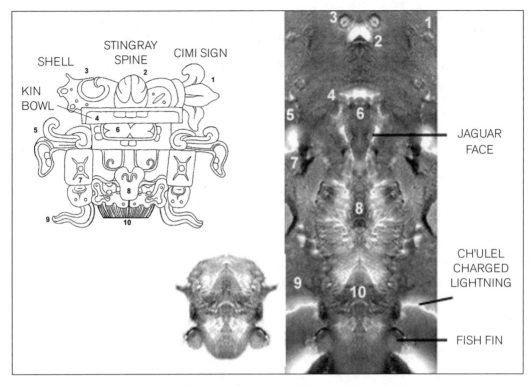

10.6 Maya First Lord and Quadrapartite Monster headdress.

LEFT, Quadrapartite headdress (found on the Lid of Pacal in Figure 3.1). Note the numbers added by the authors. The numbers match some of the common features between the two images.

RIGHT, Mirrored Mars image of Maya god First Lord (wearing Quadripartite headdress).

Drawing by George J. Haas

First Lord and the third brother, K'awil, were often depicted with cleft heads, the result of being struck with lightning charged with "ch'ulel" (life energy), which the Maya believed was present everywhere. Lightning would penetrate a mirror worn on the forehead—the third eye? This was the process by which the Divine Being imparted wisdom. This also opened a portal to the soul that allowed the transmission of energy from the Otherworld, and thus the Maya overcame death and won resurrection for their souls and for the universe as a whole.[7]

The mirrored Mars image of the Maya First Lord (Figure 10.6) depicts him with fish fins on his cheek. The ch'ulel-charged lightning is seen entering his skull and he wears the Quadripartite Monster as his headdress! We have used numbers in Figure 10.6 to point out many similar features between the Maya Quadripartite Monster and the mirrored image on Mars.

In Mayan iconography, the Quadripartite Monster is found in many composite variations. It is used as a scepter or headdress and can be found as the rear head of the two-headed Cosmic Monster. It can also be found as an independent image at the base of the World Tree.[8] When worn as a headdress, it is presented as a Cosmic Being that has no body. Its segmented features are fashioned into a composite head, which consists of a specific arrangement of symbols and glyph tags. The glyphs may vary, depending on the relative function of each headdress.

The basic set of glyphs depicts a stingray spine, an offering plate, a shell motif, and crossbands. Although these signs and glyphs may be presented in a variety of sets and groupings, their overall meaning is the same.

In the illustration of the Quadripartite Monster in Figure 10.6, the Maya depict the facial features such as eyes, nose, and mouth in a complex metaphysical manner. The forehead consists of a flat bloodletting bowl that displays the "kin" (Sun) sign. Inside the bowl rests a stingray spine, which represents the blood of the Middleworld. A spondylus shell glyph can be seen at the left, which symbolizes the water of the Underworld. On the right, the Upperworld is represented with a "cimi" or death sign and a foliage sign that reads "way,"[9] meaning to transform into a spirit companion.[10] There are also crossbands that signify the path of the Sun crossing the Milky Way.

Although the line in the center of the cimi sign is angling in the opposite direction and the foliage sign only has one leaf, the intended design seems fairly evident (Figure 10.7).

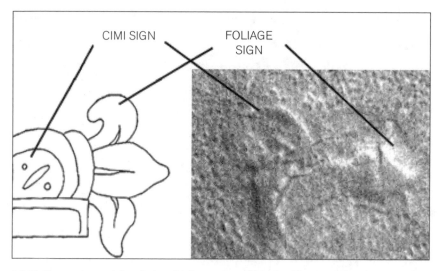

10.7 Comparison of the cimi and foliage signs from the Quadripartite Monster headdress on Mars and the feature marked 1 in the drawing in Figure 10.6. Note that the bar and dots in the cimi sign are also present.

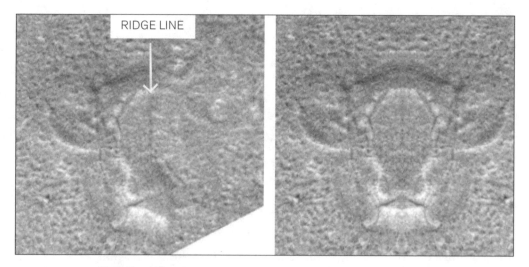

10.8 Cow's head.

LEFT, Figure 10.7, contrast-enhanced and rotated.

RIGHT, Mirrored image of cow's head.

Upon closer examination of this most peculiar structure, we found it to have an extremely intricate internal geometry. This is the same structure that Richard Hoagland referred to as "lawn furniture," after his initial evaluation of the entire strip on the night of its release on the Art Bell "Coast to Coast" radio show.[11] After the report by Hoagland, we rotated the image to try to gain a better perspective of its inner geometry. This "out of place" formation suddenly started to take on the impression of another half face. After we identified what appeared to be an eye, a large ear, and a muzzle, the probability of this complex structure being another half face became increasingly high. A mirror flip was performed to the left of the structure's defined ridge line, which was oriented vertically (Figure 10.8).

We were intrigued to see that the resulting image, when mirrored, appeared to be the head of a cow (Figure 10.8). The animal's head has a well-defined, almond-shaped ear that points outward from the side of the head; the face has an obvious eye, a muzzle, and even an elaborate crown. The cow itself has interesting symbolism. In the Vedas, the sacred text of the Hindus, the cow plays a divine and cosmic role as archetype of maternal fecundity. The cow is seen as a cloud swollen with the fertilizing rain, which falls to the Earth when the spirits of the wind, who are souls of the dead, kill the heavenly animal. They eat the cow only to bring it back to life again. Acting as the container of the water—a cloud, the cow is often given the task of guiding the souls of the dead to the heavens.

A cow was also taken to funeral pyres, where it was sacrificed and then tethered to the dead person's left foot and laid down with the flesh of a corpse. Mourners would chant hymns in an effort to raise the deceased and the cow, which would guide the soul along the Milky Way up to Heaven.[12] In a case of contour rivalry, the death sign and the spirit-companion sign of the Maya religion, when mirrored, transform into a Hindu symbol for spirit companion of the dead!

Chac-Xib-Chac and the Four Cardinal Points

When the adjacent side of First Lord and the Quadripartite Monster is mirrored along the same partition line, this west side of the swath reveals further evidence of a Mesoamerican pantheon of gods at Cydonia. The first image to arise out of the apparent chaos is Chac-Xib-Chac, the Maya god of rain (Figure 10.9). Notice the shell headdress and ear flares, the long snout, and goggle-shaped eyes. The Maya Rain god Chac was known by the Aztecs as the goggle-eyed Tlaloc. Linda Schele describes Chac-Xib-Chac as the prototype of the long-nosed god Chac, who may appear in animal form. He is distinguished by a shell crown and may also have a fish fin on his face and wear shell ear flares.[13] At times the Chac god is depicted with a serpent hanging out of his mouth, as is evident in an Early Classic period glyph seen in Figure 10.10. Amazingly, a similar serpent head juts out of the mouth of the Martian version of Chac-Xib-Chac.

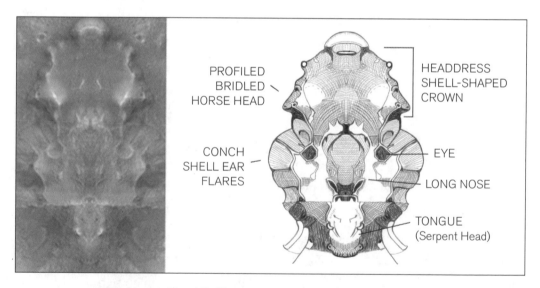

10.9 Martian Chac-Xib-Chac.
LEFT, Mirrored image (believed to represent Maya god Chac-Xib-Chac).
RIGHT, Analytical drawing by George J. Haas.

Like many of the Mesoamerican gods with their multi-faceted aspects, Chac-Xib-Chac is not just a single deity, and his attributes are complex. According to Mesoamerican mythology, Chac-Xib-Chac could transform from a single god into the four Chac gods, which designate the four cardinal points. Each of these gods is identified with the compass and is associated with one of the four directional points—with a fifth Chac nestled at the center.[14] The central Chac was called Wacah Chan ("raised-up-sky"), which was seen as the World Tree that connected all four of the directional gods running through it like a cross (Figure 10.11).

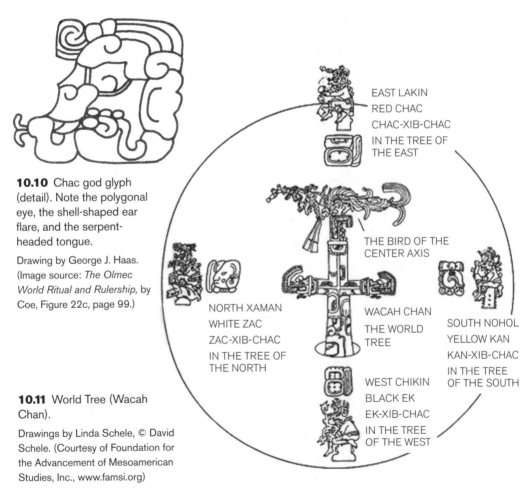

10.10 Chac god glyph (detail). Note the polygonal eye, the shell-shaped ear flare, and the serpent-headed tongue.

Drawing by George J. Haas. (Image source: *The Olmec World Ritual and Rulership*, by Coe, Figure 22c, page 99.)

10.11 World Tree (Wacah Chan).

Drawings by Linda Schele, © David Schele. (Courtesy of Foundation for the Advancement of Mesoamerican Studies, Inc., www.famsi.org)

EAST LAKIN
RED CHAC
CHAC-XIB-CHAC
IN THE TREE OF
THE EAST

THE BIRD OF THE
CENTER AXIS

NORTH XAMAN
WHITE ZAC
ZAC-XIB-CHAC
IN THE TREE OF
THE NORTH

WACAH CHAN
THE WORLD
TREE

SOUTH NOHOL
YELLOW KAN
KAN-XIB-CHAC
IN THE TREE
OF THE SOUTH

WEST CHIKIN
BLACK EK
EK-XIB-CHAC
IN THE TREE
OF THE WEST

The Maya cosmos comprised a universe made up of three layers through which the World Tree passed, layer by layer. The Upperworld was of the heavens, the Middleworld the earth, and the Underworld the dark waters that lurked below.[15] The four directional Chacs were each associated with their own vital branch of this central tree, along with a bird and various symbolic colors. The eastern quadrant of this sacred space was the red Chac, called Chac-Xib-Chac. The northern quadrant was the white Chac, called Zac-Xib-Chac. The western region was seen as the black Chac, called Ek-Xib-Chac, and the south was the yellow Chac, called Kan-Xib-Chac.[16] The arrangement of the gods does not fit into the normal, accepted order of compass points because in the Maya conception of the cosmos, it was the east and not the north that was placed at the top of a map.[17]

Because the Chac god was principally seen as the Rain god, the basic colors of these four directional gods have been compared to the significance of the four cardinal humors of medieval times by scholars.[18] The four humors are considered to be four liquids that are responsible for our medical and mental well-being. The colors of these bodily secretions are: red for blood, which is the essence of life; white for the respiratory discharge of phlegm; yellow for bile or for choler, which was considered the source of anger; and black bile that signified the state of melancholy.[19]

Taking a closer look at the Mars image of Chac-Xib-Chac, we noticed that the overall portrait of this Rain god was actually a composite image that also included the four other Chac gods within his own face. The four additional Chac heads were discovered stacked one within the other like a Russian matrushka doll. Each of these four hidden Chac gods repeated a variant portrayal of a humanoid rider and horse head (Figure 10.12).

These four images within the main Chac god portrait bring to mind the Four Horsemen of the Apocalypse. In this context it has been suggested that the horse was associated with cosmic forces.[20] To support this assignment

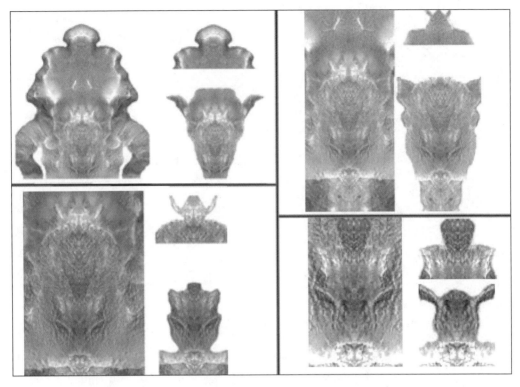

10.12 The four Chac gods on Mars (the Four Horsemen of the Apocalypse). Note that these horsemen are all found in Figure 10.9.

of the horse as a signifier of the four cardinal points, we look to an ancient mosaic found in the Middle East. This rare and highly complex floor mosaic, found in the Holy Land in Israel's Jezreel Valley, depicts a circular zodiac, including the twelve constellations, within an outer cosmic square.

At each of the four opposing corners of this mosaic rests a separate winged deity (not visible in the illustration) that marks the four cardinal points. Similar birds accompany each of the four directions in an Aztec version now called the Codex Fejérváry-Mayer. Within a sphere at the center of this ancient zodiac found in Israel is a fifth deity (like Wacah Chan of the Maya) harnessing four horses (Figure 10.13).[21]

10.13 Byzantine floor mosaic (Beth Aipha, Jerusalem, sixth century A.D.). Note the four horses.

Drawing by George J. Haas. (Image source: *Atlas of the Jewish World*, by de Lange, page 89.)

This central figure, with the halo-like crown, is seen here as Yahweh, the personification of the Sun or Helios.[22] Also notice how the highly stylized mask-like presentation of the horse heads in the floor mosaic in Figure 10.13 resembles the masks of the four horse heads seen in the Martian representation of the Chac gods in Figure 10.12.

In the Revelation story in the Bible, each of the Four Horsemen rides a horse of a different color, just as the four Chac gods of Mesoamerica are associated with their own directional color. The first of the four horses of the Apocalypse is represented as white, just as the Chac of the north is white. The second horse is red, just as the Chac of the east is red. The third horse is black, just as the Chac of the west is black, and the fourth horse is pale while the fourth Chac is yellow, signifying the south. The similarities between the four Chac gods of the Maya and the Four Horsemen of the Apocalypse are unmistakable.

A further relationship between the horse and the agricultural attributes of the Rain god (in this case, the Chac god) comes from the allegorical stories of Greek and Roman mythology. The mythological winged horse known as Pegasus was thought to have the ability to find underground springs. This was achieved by the horse digging and striking the ground, causing wells to spring forth, just as he did with the Fountain of Hippocrene.[23] Pegasus was seen not only as the equivalent of a water god, but also as the embodiment of the Rain god himself.[24]

Furthermore, just as we have seen with the Maya, each of the four Chac gods is not only associated with a color; it is also presented with a particular bird. In a celestial context, this same idea of a winged horse is associated with the symbolism of a bird.[25] We found that in ancient Rome it was the horse that was annually sacrificed to Mars to ensure the yearly harvest.[26] The severed head of a horse was then hung on the gates of the city to further appease this unreasonable god we associate with the Red Planet.

The pairing of Chac-Xib-Chac with First Lord's Quadripartite Monster headdress on this immense structure on Mars could not be more fitting, as quadripartite means having four parts. Furthermore, according to scholars, the analogous characteristics and features of First Lord and Chac-Xib-Chac are interrelated.[27] Here again we find two companion Maya gods that are at times seen as one. When Chac-Xib-Chac is associated with the planet Venus, as the evening star, he is then also identified as First Lord.[28] Incredibly, as if all this encoded imagery in one composite structure on Mars were not enough, there was more to be found!

The Bridled Horse

Along the shell headdress of the Chac god on the Martian structure, there exists a feature that Hoagland has compared to a Giza-scale pyramid (Figure 10.14). However, when the left side of this structure is perceived as a product of contour rivalry, the elbow formation of the first Horseman takes on the profile of a bridled horse's head (Figure 10.14).

Notice the horse's muzzle, eye, ear, and bridle. The presence of a bridle with all its trappings signifies not only a rider; it can also be interpreted as man's control over nature. Symbolically, a bridle is an instrument of restraint that harnesses the spirit of the self.[29] Could this bridled horse head found at Cydonia signify the control and management of the four cardinal points by the Chac god?

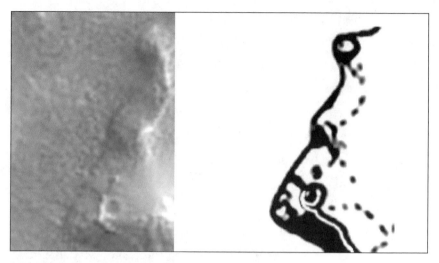

10.14 Horse head profile (Hoagland's Giza-scale pyramid).

LEFT, Bridled horse head profile (cropped from the image of Chac-Xib-Chac in Figure 10.9).

RIGHT, Analytical drawing by George J. Haas. Note the bridle ring and straps on the horse's nose.

The Fresco at Tulum

Contained within the isolated walls of a sadly deteriorating temple along the coast of Quintana Roo in the Yucatan, we found a rarely publicized mural that captures the Chac god riding a most unusual animal. The temple is a strategically located complex that sits high on a cliff of an ancient Mayapan town called Tulum, overlooking the Caribbean Sea. Preserved on the walls of the temple, which has been titled Structure 16, is a highly controversial mural that depicts the Chac god riding an animal that has been casually described as a horse (Figure 10.15).

Some critics have argued that the Chac god is actually riding a jaguar. However, this so-called jaguar depicted on the Tulum fresco has hooves and a flowing tail. According to archaeologist Arthur G. Miller, although some structures at Tulum and the surrounding areas may date to before

1250 A.D., the murals found in Structure 16 date to the Middle Post-Classic period, to about 1400 A.D.[30] That would place the creation of the fresco at Tulum well over a hundred years before the Spaniards reintroduced horses in the Americas.

So how do scholars explain this out-of-place equine at Tulum? In his book *On the Edge of the Sea*, Miller cautiously suggests that the Chac god is "riding an unidentifiable quadruped" but he never fully addresses its implications.[31] In his book *The Maya*, Michael D. Coe not only confirms that the Chac god is indeed riding a quadruped; he identifies the animal as a horse.[32] The renowned archaeologist J. Eric S. Thompson, in his 1954 book *The Rise and Fall of Maya Civilization*, also acknowledges the presence of this amazing image of the Chac god mounted on a horse.[33] According to the scholars who have studied the evolution of the horse, the animal disappeared from the area over 8,000 years

10.15 Maya Chac god riding a horse (Tulum fresco, detail of restored mural, 1400 A.D.). Note that, although this is a damaged area of fresco, it is obvious that the Chac god is astride a four-legged, hoofed beast that has been identified as a horse.

Drawing by George J. Haas. (Image source: *Ancient Past of Mexico*, by Reed, page 338.)

ago.[34] However, current evidence supports the stance that the Maya not only produced pre-Conquest sculptures of horses;[35] they also envisioned one of their gods riding one. Perhaps it is no coincidence that the Tulum horse is similar in stature to the ones that were roaming the country after the last ice age.

Ever since its discovery in the late 1840s by Stephens and the identification of this incredible fresco with that of a highly stylized horse, its origins have been locked in deep controversy. Upon visiting Tulum to investigate this painting for ourselves in both 2001 and 2003, the authors found the doors of this temple closed to the public. Objectively, we find it amazing to discover that not only is a horse depicted on this Mesoamerican fresco, but of all the gods the Mayapan artist could have chosen to place on a horse, it's the Chac god who is astride this four-legged beast. In light of what we have seen on Mars of this kinship between the horse and Chac god, we believe that this historical symbolism and parallel iconography is much more than a coincidence.

The Lord K'awil

The Maya divided the progression of time into four quadrants of 819 days each. Just as each of the four cardinal directions was presided over by a different Chac god, each quadrant of time was presided over by a K'awil god. As with the Chac gods, each K'awil god had his own quadrant and color, matching those of the Chac gods: red-east, white-north, black-west, and yellow-south.[36]

According to the mythology of the Maya, the Creator Couple—the First Father and First Mother—gave birth to twins, First Lord (Venus) and First Jaguar (Sun). They also had a third son, K'awil. Today, many Mayanists believe that these three gods were just different aspects of one divine being. To further this argument, we take another look at the eastern-side split of the partition, which contains the image of First Lord wearing the Quadripartite headdress. When one looks down onto this structure and views the chest area of First Lord, it appears that the third brother, K'awil, is actually portrayed in this same image (Figure 10.16). This additional composite element makes the triad complete, because the Quadripartite Monster (worn as the headdress of First Lord) represents the Sun (First Jaguar).

As with First Lord, K'awil was also struck with the ch'ulel-charged lightning, but it entered through the smoking torch/axe head, which penetrated a mirror in his forehead.[37] The god K'awil is described by Schele and Freidel as follows:

... he is always zoomorphic in aspect. His most important feature is a smoking object—such as a cigar, torch holder, or ax head—which penetrates a mirror in his forehead.... His face always has the zoomorphic snout traditionally called a long-nose.[38]

K'awil also represented blood, semen, and corn, which are aspects of the Celestial Dragon or Cosmic Serpent. Thus he symbolizes its presence in the human world.[39]

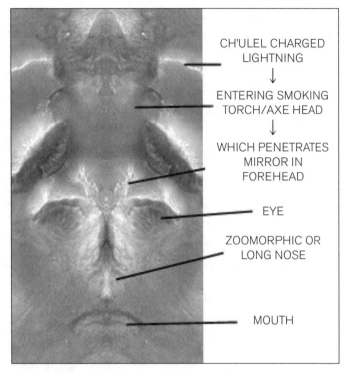

CH'ULEL CHARGED LIGHTNING
↓
ENTERING SMOKING TORCH/AXE HEAD
↓
WHICH PENETRATES MIRROR IN FOREHEAD

EYE

ZOOMORPHIC OR LONG NOSE

MOUTH

10.16 Maya god K'awil at Cydonia. The image of the Maya First Lord also forms the image of K'awil.

The World Tree

In the discussion of the Chac gods, we mentioned that the World Tree, called Wacah Chan, is at the center of the four quadrants (Figure 10.17). It is depicted in the form of a cross and is marked with the Maya symbol of divinity, the tzuk or partition head. This partition head announces it as a divine or holy thing. It is at this central axis of the world that the Creator god, known as First Father to the Maya, communicates between the natural and supernatural worlds. According to Schele and Freidel:

> The bejeweled, square-snouted serpents which usually terminate its branches represent flows of liquid offering—human blood.... The

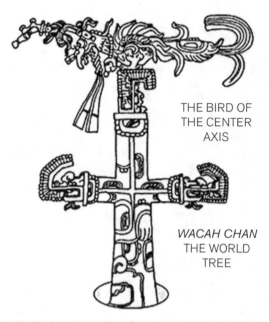

THE BIRD OF
THE CENTER
AXIS

WACAH CHAN
THE WORLD
TREE

10.17 Maya World Tree (Wacah Chan).

Drawings by Linda Schele, © David Schele. (Courtesy of Foundation for the Advancement of Mesoamerican Studies, Inc., www.famsi.org)

Tree is the path of communication between the natural and supernatural worlds as it is defined at the center of the cosmos.[40] ... The act of communication between the human world and the Otherworld was represented by the most profound symbols of Maya kingship: the Vision Serpent and the Double-headed Serpent Bar.[41]

When the opposite split of the Quadripartite headdress worn by the Maya First Lord was mirrored, it produced the image of Chac-Xib-Chac and the four Chac gods, which could also be interpreted as the Four Horsemen of the Apocalypse. When the opposite split of the Maya god K'awil is mirrored, the resulting image is visually stunning (Figure 10.18). In *The Code of Kings,* Schele and Mathews describe the Maya World Tree as follows:

A great tree emerges from the bowl of sacrifice and rises behind the body of the dying king. The trunk carries a "tzuk" head (partition head) to mark it as the center partition. Mirror signs define its substance as something "shiny," while te', "tree," signs assure us that it is a tree. The branches terminate in beaded flowers with square-nosed, bejeweled serpents emerging from their centers. These are the personified stamens of the blossom. Ceiba trees flower in late January and early February, just at the time Maya myth says First Father raised this tree. Today we know it represents the Milky Way as it stretches across the sky from the southern horizon to the north. The White-

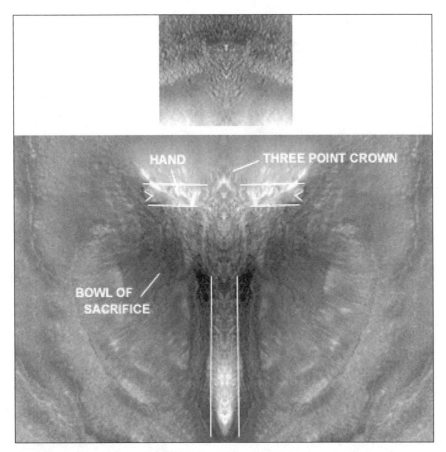

10.18 Mirrored image of Maya World Tree on Mars (World Tree with dying king).
TOP, Maya First Mother? (detail of World Tree on Mars).
BOTTOM, Highlighted Maya World Tree on Mars, with possible image
of First Mother, above.

Bone Snake at the base of this image represents the hole in the southern horizon that is the passageway of the souls and ancestors who
have been reborn. The Maya name for the Milky Way was Sak Beh, the
"White Road."[42]

After all the symbolic imagery we have seen thus far, we thought it would
be difficult to find another image that would truly startle us. However, con-

sidering the parallels to the mythology of the Christian religion and the crucifixion of Jesus Christ, our next discovery truly did. The image in Figure 10.18 appears to show a spectacular crucifixion scene in which a cross-shaped World Tree appears to rise from the bowl of sacrifice behind the central figure. It also appears that the arms of the World Tree/Crucifix are spread across the bosom of an image that could only be interpreted, in this context, as that of the First Mother of Maya mythology.

The Womb Tomb

When the image of the World Tree/Crucifix is inverted (turned upside down), we see another amazing example of contour rivalry, which we have seen so often in these Martian geoglyphs. The inversion produces an image that is a remarkable analogy to the Maya Womb Tomb, which enables the spirit to be reborn (Figure 10.19). Mythologist Douglas Gillette explains this idea of the Womb Tomb as follows:

> Another image of blood, descent, and rebirth that scholars of comparative religion and mythology have discovered is the universal idea of the so-called womb-tomb. This idea, which appears in the myths of all ancient agricultural peoples, sees the Underworld as the womb of the Earth. The dead body is placed in the tomb, often in a fetal position, and painted red to symbolize the blood of the womb with its life-generating power. When the time is ripe, the soul is reborn from the womb-tomb and rises as a god-like spirit into eternal life.[43]

As we analyze the western side of this bifurcated image, we find a figure that appears to be a body in a coffin-shaped grave or tomb. The shape of the coffin extends in such a manner that it projects a phallic appearance, or an arrow pointing upwards (Figure 10.20A). The coffin appears to be ascending from what could be perceived as a womb. Notice the figure's head, the crossed arms holding what appear to be doves, and a mask-shaped

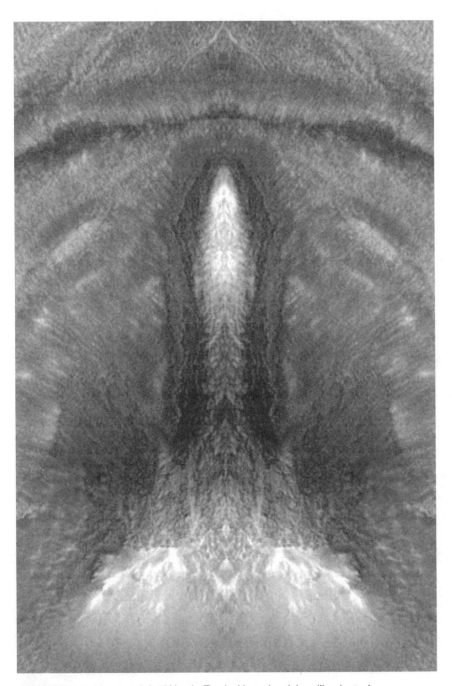

10.19 Mirrored image of the Womb Tomb. Note the rising, illuminated, crystallized figure.

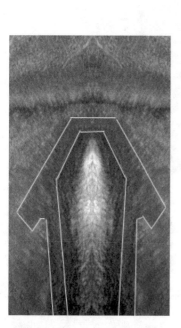

10.20A Mirrored image of the resurrection of body in a tomb. Note the coffin-shaped grave or tomb.

10.20B Body in a Tomb (analytical drawing). Note the five-pointed star headdress, the crossed arms holding doves(?), and the mask-shaped apron.

Drawing by George J. Haas

10.20C Maya Stela P (Copan). Note the folded arms (with outward twisted hands that echo the shape of the doves(?) in the Body in a Tomb) and the elaborate apron, including a small mask.

Drawing by George J. Haas. (Image source: *Scribes, Warriors and Kings*, by Fash, Figure 50, page 98.)

apron. The figure also has a star-shaped crest above his head (Figure 10.20B). As for our speculation on the significance of the star, we quote Cirlot's *Dictionary of Symbols:*

> [T]he star is a symbol of the spirit. Bayley has pointed out, however, that the star very rarely carries a single meaning—it nearly always alludes to multiplicity. In which case it stands for the forces of the spirit struggling against the forces of darkness. This is a meaning which has been incorporated into emblematic art all over the world. For this reason, "identification with the star" is possible only to the chosen few.... The five-pointed star is the most common. As far back as in the days of Egyptian hieroglyphics it signified "rising upwards towards the point of origin."[44]

The Maya Stela P from Copan (Figure 10.20C) shows a figure that is amazingly similar to the body in the tomb on Mars. The position of the arms and the elaborate apron below the waist are very reminiscent of the image in Figure 10.20A on Mars. Notice the position of the crossed arms. The outward-twisted hands echo the shape of the dove forms seen within the Figure in a Tomb. Also note the elaborate apron, including a small mask. These monolithic structures were created by the Maya to display the king in the guise of the World Tree.[45]

On the obverse side of Stela P stands the complete image of an elaborately outfitted king, who becomes the human embodiment of the World Tree. As the World Tree, the king transforms into a pillar-like manifestation of the central axis of the world.[46] At the center of this axis is a passageway at the base of the White-Bone Snake (the Milky Way), where souls are reborn.[47]

With this in mind, it can be presumed that the Figure in a Tomb (on Mars) is another manifestation of the World Tree, much in the same manner as those seen on Maya stelae. Another intriguing observation is that

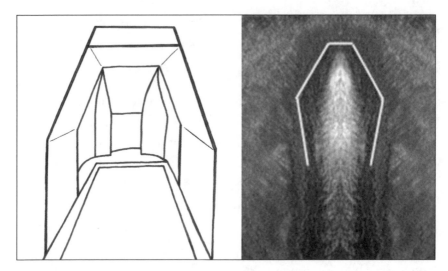

10.21 Shape comparison of Pacal's tomb and Mars tomb.
LEFT, Maya King Pacal's tomb (with the shape highlighted).
Drawing by George J. Haas
RIGHT, Mars tomb (with the shape highlighted in white).

the glimmering Figure in a Tomb projects the feeling of rising or levitation. The cascade of the segmented elements surrounding the Figure in a Tomb gives the appearance that he is being resurrected in a crystalline beam of light.

In evaluating the outer shape of the coffin in the Mars image, we were reminded of the tomb of the great Maya king Pacal, which included his sarcophagus. Notice the shape of Pacal's tomb in comparison to the shape of the coffin or grave in the Mars image illustrated in Figure 10.21.

This idea of a king being transformed into both a tree and a stela and then being resurrected is also seen in Egyptian mythology. The correlation between the two cultures begins when the Egyptian god Set (Seth) tricks his brother Osiris into climbing into a sarcophagus. Set closes the lid and the sealed casket is then cast into the Nile (which is seen here as the Milky Way, or White-Bone Snake). The sarcophagus eventually washes ashore to rest in a thicket along the coast of Byblos. A beautiful tree quickly grows up, incorporating the sarcophagus in its trunk, much like the Maya idea of the king transforming into the World Tree.

In the Egyptian story, the King of Byblos discovers this beautiful tree and has it cut down and carved into a central pillar to adorn his palace. This carved pillar is similar in form and symbolism to the Maya stela, which encapsulates the king in its tree form. When Isis finally discovers the location of the pillar, she has it cut open. It reveals the mummified corpse of her husband, Osiris.[48]

This concept of a corpse being equated with a living tree is also found in the traditions of South America. Just as the Cydonia image of the World Tree transforms into a Figure in a Tomb, the Inca equate the mummies of their ancestors with trees and addressed them both by the same name, Malqui.[49]

The Principal Bird Deity

The next thing we noticed was that the Figure in a Tomb appears to be emanating from an angelic or god-like being at the base of the coffin (Figure 10.22). We were familiar with many winged heavenly beings, eagle men, and bird men depicted in ancient cultures from Sumer, Egypt, Asia, and Mesoamerica. We found similar winged deities carved at the base of many Maya stelae that feature kings in the guise of gods. Figures 6.11 and 6.13 in Chapter 6 presented a comparison of the winged deity from Assyria and a winged Quetzalcoatl from Mesoamerica.

The Principal Bird deity was found at the top of the Maya World Tree and that is where this figure is found on the Mars World Tree, inverted in Figure 10.19. Is this angelic figure at the base of the Figure in a Tomb on Mars the equivalent to what the Maya referred to as the Principal Bird

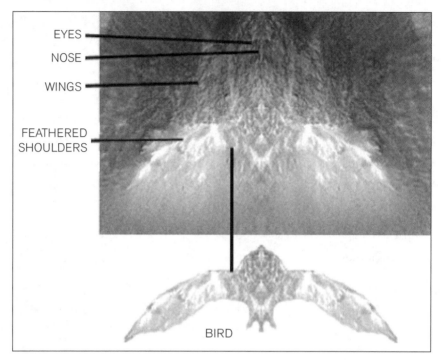

EYES
NOSE
WINGS
FEATHERED
SHOULDERS
BIRD

10.22 Martian Principle Bird deity.
TOP, In humanoid form.
BOTTOM, In animal form.

deity? In another example of finding images within images on the Martian surface, we may find the answer by looking at the composite nature of this bird-faced structure.

The comparative image in Figure 10.23 shows a typical illustration of the Principal Bird deity found on a Maya temple at Tikal, above a similar image comprising the neck and shoulders of the winged deity as presented in Figure 10.22. Notice how the glimmering wings appear to burst open from the sides of the avian head. A small head can be seen within the crown area, with the signature "tri-leaf" headdress, in both the Martian and Mayan images. Two perfectly round eyes with tiny black pupils are also common to both images (Figure 10.23C).

10.23A Principle Bird deity (from Mayan temple).

Drawing by George J. Haas. (Image source: *Maya Hieroglyphic Writing,*
by Thompson, 1971, Figure 20, #10.)

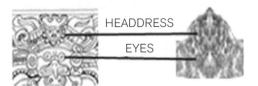

10.23B Mirrored bird image on Mars. Note
the similar faces found in the headdresses, the
three-point crown, and the perfectly circular
eyes of the birds.

10.23C Comparison of the heads.

The Lords of Xibalba (the Underworld)

The idea of a common seed of all religions is not a new idea. Scholars for
years have noticed the similarities in religious beliefs, symbols, and teachings of various cultures throughout the world. If indeed the world's religions evolved from a common source, they have ultimately deviated from
each other in many ways. We have noticed that most of the religions of
the world believe in a dualism in divinity, as opposed to the duality which
was at the very heart of Maya culture and beliefs. Dualism is the belief that
the forces of good and evil are irreconcilable enemies. This conflict is seen
in such oppositions as Christ and Satan.[50]

The Maya belief in the dualistic nature of the Divine Being is that the seed of all things good and evil are within the Divine Being, and therefore all things of Creation carry the seeds of their own destruction.[51] These dualities are in constant exchange; one ultimately leads to the other. The Maya believed that from life comes death, which ultimately leads to new life. This idea is even expressed with the simple passing of the day, where from light there is darkness, and once again there is light.

This belief was at the core of the Maya way of life and was expressed in the depiction of their gods. God was not only the ultimate Divine Being; God was also a twin, to represent the duality within, which expressed man/woman, good/evil, life/death, and so on. We believe we have found further expression of this idea in another incredible display of contour rivalry in this geoglyphic mountain on Mars.

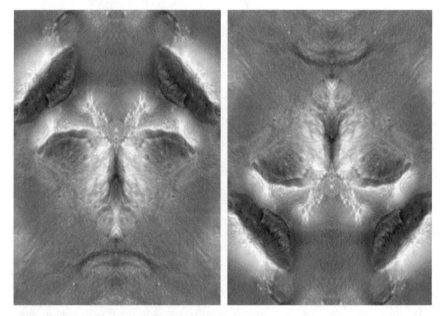

10.24 Context image. K'awil and inversion.
LEFT, Mirrored image of Maya god K'awil (from Figure 10.16).
RIGHT, Same image inverted.

When we view the inverted image of K'awil from Figure 10.16 (Figure 10.24), more stunning images appear. If we take a closer look at this image, it unfolds as a composite of two images. The first has an undeniable resemblance to a classic depiction of the Judeo-Christian-Moslem demon known as Satan, who appears to be peering down into the flames of Hell (Figure 10.25). His horned head is found just below the nose of K'awil. Below him, right between the "eyebrows" of K'awil, is the Martian equivalent of a Hindu demon god known as Mahishasura (Figure 10.25).

A cape cloaks the Satan-like figure and he wears a medallion on his chest. His open cape exposes flames erupting from a dark abyss. This horned Satan-like image (Figure 10.26) appears similar to an ancient Maya god known as Lord Death, acquired by Maurice Cotterell from the sarcophagus Lid of Pacal in 1992. For our study, anomaly hunter Jim Miller has provided an analogy of Cotterell's image of Lord Death (Figure 10.27).

When recreating a transparent overlay of this image for the publication of this book, Miller discovered that the rotation required to reproduce Cotterell's image of Lord Death was 72 degrees. The pivotal number 72 is significant here, because it is this exact number that denotes

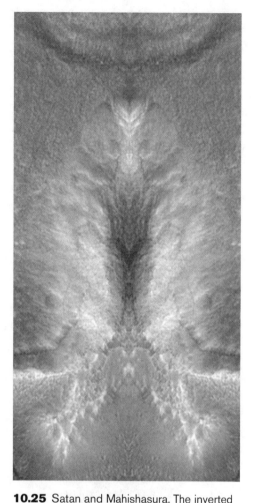

10.25 Satan and Mahishasura. The inverted image of Maya god K'awil becomes:
TOP, Satan-like demon.
LOWER, Hindu demon Mahishasura.
Notice how the Satan-like demon appears to be looking down into a flaming abyss, while the shoulders and neck of the demon Mahishasura also appear to be in flames.

HORNS

10.26 Satan-like demon (cropped from Figure 10.25). Note the horns.

10.27 Lord Death (Lid of Pacal overlay). Note the horns. Overlay and colorizing by Jim Miller, after Cotterell.

the act of precession from the Pole star. This celestial alignment is directly related to the Maya idea of the partition of the sky, at the time of Creation, into the four cardinal points. It is at this exact axis mundi that this image of Lord Death is found at Cydonia.

Below the image of the Satan-like demon a second demon can be seen (Figure 10.25). Flames emanating from his shoulders form a perfect contrast to the "feathered" shoulders of his opposite companion, the Winged Deity seen in Figure 10.22. The illustration in Figure 10.28 features a detail of a Hindu demon sitting astride the prow of a royal barge in Thailand. A comparison of the mirrored Mars demon and the Thai demon speaks for itself (Figure 10.28).

We are able to confirm the identity of this Martian demon in the same manner as we did with the Principal Bird deity, by the images found on the neck and chest. In the neck and chest area of the demon we found a composite image of a buffalo whose muzzle has the image of a lion within it

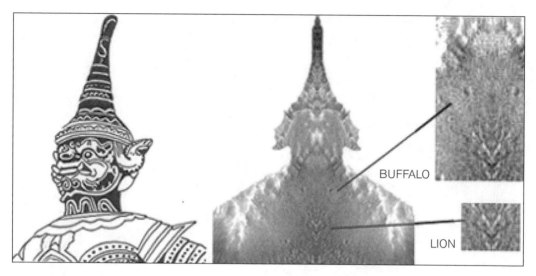

10.28 Demon comparison.

LEFT, Hindu demon (Mahishasura).

Drawing by George J. Haas. (Image and quote source: *Mind Alive Encyclopedia: Early Civilization*, 1968, page 121.)

CENTER, Mars demon.

RIGHT, Buffalo/lion totem. Above: buffalo. Below: lion. Note that the buffalo and lion images are found within the chest of the Mars demon.

(Figure 10.28). According to Hindu mythology these two beasts, the buffalo and the lion, are animal manifestations of the demon Mahishasura.

The following is an excerpt from an article titled "The Death of Mahishasura":

> Mahish, the king of the demons and usurper of the throne of heaven, was shocked and enraged by the disastrous events on the battlefield. He reverted to his own form, a "buffalo," and charged about on the battlefield. He ran wildly at Durga's divine soldiers, goring many, biting others and all the while thrashing with his long, whip-like tail. Durga's lion, angered by the presence of the demon-buffalo, attacked him. While he was thus engaged, Durga threw her noose around his neck. To escape this trap, Mahishasura discarded the buffalo and

assumed the form of a "lion." Durga beheaded the lion, and the demon escaped in the form of a man.[52]

The Mars image produces not only a likeness to the "usurper of the throne of heaven" of the Judeo-Christian-Moslem religions—Satan—but also the "usurper of the throne of heaven" of the Hindu religion—Mahisha-sura! The complexity of this composite glyph does not end there, as it also displays the transformation of Mahishasura into the buffalo and the lion!

Overworld and Underworld Mirror Reflections

As mentioned earlier, the Maya divided the world into three planes of existence: the Overworld, the Underworld, and the Earth plane. Modern theologians have described the three-layered structure of the universe as it has emerged from the Elsewhere and Elsewhen of the Otherworld as "a place where an eternal now exists—a time-space forever repeating its sacred patterns without beginning and without end."[53]

The Maya belief in the duality within the Divine Being led to their regarding the Overworld and the Underworld as "mirror reflections" of one another. Thus the demons of the Underworld were actually the mirror reflections of the gods of the Overworld.[54] Finding these key gods and demons from other major religions of the world among all the Maya images was a great quandary. How did these alternative gods tie into the Mesoamerican pantheon of gods that up to this point had been so consistent? Again, it was a passage from Douglas Gillette's book that put the pieces together (Figure 10.29):

> In the Maya Creation text, the *Popol Vuh* [it states], "First Father, known in this text as One One Lord, had a twin himself—Seven Lord. It was this first pair of twins that the Lords of Xibalba (the Underworld) sacrificed, and First Lord and First Jaguar brought back to life."

The Maya saw the first set of twins and their sons as ultimately one and the same being—the husband/son of First Mother.

In the *Popol Vuh,* One One Lord and Seven Lord confront and are defeated by their archenemies, the demon kings One and Seven Death. We know from the parallels in their names that these two sets of twins were twins themselves—*Overworld and Underworld mirror reflections of each other* [emphasis added].[55]

Symbolist J. E. Cirlot says the mandorla is used to symbolize the Upper and Lower Worlds of Heaven and Earth. The union of the two worlds or the zone of intersection and interpenetration is represented by an almond-

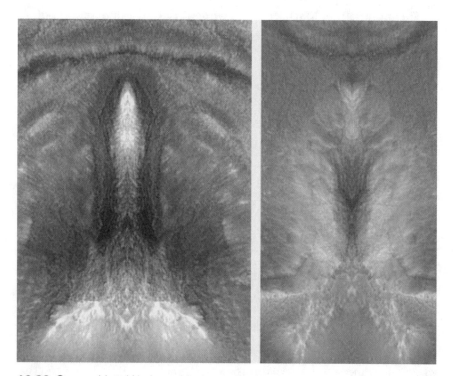

10.29 Overworld and Underworld mirror reflections.

LEFT, Image of One One Lord and Seven Lord of the Overworld (with their mirror reflections on the right).

RIGHT, One and Seven Death of the Underworld.

shaped shield (Figure 10.30) that forms from the intersection of two circles. Cirlot goes on to state:

> The mandorla, like the twin peaked mountain of Mars, embraces the opposing poles of all dualism. Hence it is a symbol also of the perpetual sacrifice that regenerates creative force through the dual streams of ascent and descent (appearance and disappearance, life and death, evolution and involution).[56]

Cirlot relies on his mentor, the German scholar Marius Schneider,[57] to clarify this idea of the mandorla. Again we use the quote that we have used extensively in this book:

> The mountain of Mars (or Janus) which rises up as a mandorla of Gemini is the locale of Inversion—the mountain of death and resurrection; the mandorla is a sign of Inversion and interlinking, for it is formed by the intersection of the circle of the Earth with the circle of heaven. This mountain has two peaks, and every symbol or sign alluding to this "situation of Inversion" is marked by duality or by twin heads.[58]

How could Schneider, writing in the 1930s and 1940s, have been so prophetic about the revelations that would occur at this bifurcated mountain (Figure 10.31) on Mars in 1998? Were these his words, or was he quoting from an ancient text, now lost? Did Schneider have access to profound information that was possibly discovered by Nazi archaeologists in Central America during his military excursions throughout World War II? We may never know.

According to the Maya, this arching mandorla and its intersecting points where the Upper and Lower Worlds meet, through the World Tree and the partitioning of Heaven and Earth, guarantee the renewal of life through sacrifice.[59] This sacred intercommunication between man and these two

10.30 The mandorla. Note the mandorla symbolizes the intersection of the two circles of Heaven and Earth.

Drawing by George J. Haas. (Image source: *Dictionary of Symbols,* by Cirlot, 1996, page 203.)

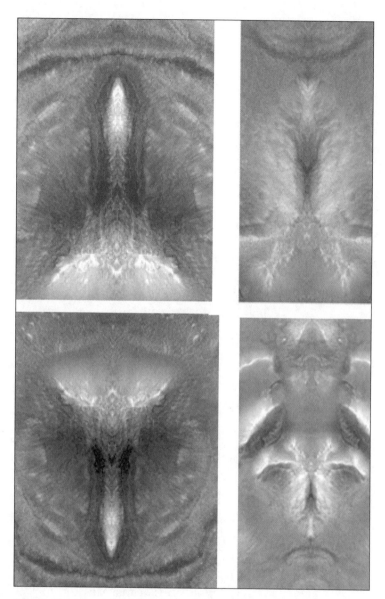

10.31 The Mountain of Death and Resurrection. Note each of these figures is the opposite of the one beside it and the inversion of the one above or below it. Each image also displays twin heads.

265

worlds can only be preserved through a "rigid law [that] demands a death for each life."[60]

According to Schneider, it's the red planet we call Mars that represents this "perennial incarnation" of our primordial necessity for sacrifice and war and the shedding of human blood.[61] For, as we saw in the previous chapter, it is our very blood that completes this union. In the following chapter, this reference to the shedding of blood—our vessel of DNA—will be revealed once again in the elusive feline side of the Face on Mars.

Notes

1. David Freidel, Linda Schele, and Joy Parker, *Maya Cosmos: Three Thousand Years on the Shaman's Path* (New York: Quill, 1993), 113.

2. Ibid., 73.

3. Ibid., 140.

4. Ibid., 420.

5. Ibid., 140.

6. Linda Schele and David Freidel, *A Forest of Kings: The Untold Story of the Ancient Maya* (New York: Quill, 1990), 413–414.

7. Douglas Gillette, *The Shaman's Secret: The Lost Resurrection Teachings of the Ancient Maya* (New York: Bantam, 1997), 68.

8. Linda Schele and David Freidel, op. cit., 414.

9. Linda Schele and Peter Mathews, *The Code of Kings: The Language of Seven Sacred Maya Temples and Tombs* (New York: Touchstone, 1999), 113.

10. Personal communication from David Freidel at the Maya Weekend, University of Pennsylvania Museum of Archaeology and Anthropology, April 1, 2000.

11. The Art Bell "Coast to Coast" radio show, April 14, 1998. On the night the second Cydonia swath was released, Richard Hoagland noticed a small area consisting of odd geometric shapes, toward the lower portion of the second Cydonia strip. They appeared to be so out of place, he

whimsically titled the anomaly "lawn furniture." To explain the humorous title, Hoagland related a brief story behind this little structure to a statement made to him by none other than Dr. Michael Malin. During a 1988 interview in the *New York Times,* Hoagland recalled, "He [Malin] wouldn't believe there were artificial structures on Mars until he could see the lawn furniture." So, with this statement in mind, throughout his evaluation of Martian anomalies, Hoagland maintained the idea that he could only find indisputable evidence of ancient artifacts on Mars if he looked for the devil in the details. As stated, Hoagland believed that he had just found such evidence. He basically declared this little structure to be so unnatural that it could easily be seen as the Martian equivalent of a grandiose patio complex complete with lawn furniture. Hoagland's initial analysis of this odd structure was that it appeared to be "the most blatant feature of things that should not be down there ... but, they are!"

12. Jean Chevalier and Alain Gheerbrant, *A Dictionary of Symbols* (New York: Penguin Books, 1996), 238.

13. Linda Schele and David Freidel, op. cit., 408.

14. George C. Vaillant, *Aztecs of Mexico: Origins, Rise and Fall of the Aztec Nation* (Garden City, NY: Doubleday, 1941), 171.

15. Linda Schele and David Freidel, op. cit., 66.

16. Ibid., 66–67.

17. Ibid., 66.

18. Irene Nicholson, *Mythology of the Americas* (London: Hamlyn, 1970), 223.

19. *Webster's New World Dictionary with Student Handbook* (Nashville: The Southwestern Company, 1974), 364.

20. J. E. Cirlot, *A Dictionary of Symbols* (New York: Barnes & Noble, 1995), 152.

21. Zecharia Sitchin, *The Cosmic Code: Book VI of The Earth Chronicles* (New York: Avon, 1998), 36–37. See also J. Glen Taylor, "Was Yahweh Worshiped as the Sun?" *Biblical Archaeology Review,* May/June, 20, No. 3 (1994), 61.

22. J. Glen Taylor, op. cit.

23. Jean Chevalier and Alain Gheerbrant, op. cit., 522.

24. Ibid., 522, 525.

25. Ibid., 523.

26. Ibid., 524.

27. Linda Schele and Mary Ellen Miller, *The Blood of Kings: Dynasty and Ritual in Maya Art* (New York: George Braziller, 1985), 60.

28. Ibid., 312.

29. Jean Chevalier and Alain Gheerbrant, op. cit., 524.

30. Arthur G. Miller, *On the Edge of the Sea: Mural Paintings at Tancah–Tulum, Quintana Roo, Mexico* (Washington, D.C.: Dumbarton Oaks, 1982), 11, 14. The fresco at Tulum is dated to a pre-Conquest date of just before 1400 A.D. If accurate, that would place the painting of the Chac god on horseback to well over 100 years before the Spanish ever arrived in the Yucatan.

31. Ibid., 55.

32. Michael D. Coe, *The Maya,* 6th ed. (New York: Thames & Hudson, 1999), 187.

33. J. Eric S. Thompson, *The Rise and Fall of Maya Civilization* (Norman, Okla.: University of Oklahoma Press, 1966), 269.

34. Elwyn Hartley Edwards, *The Encyclopedia of the Horse* (New York: Dorling Kindersley, 1994), 13.

35. In a 1974 excavation at Tancah, a site right next to Tulum, a small coral horse head was recovered among the ruins of an ancient building dated to 1350 A.D. Arthur G. Miller, op. cit., 78.

36. Linda Schele and David Freidel, op. cit., 78.

37. Douglas Gillette, op. cit., 66.

38. Linda Schele and David Freidel, op. cit., 414.

39. Peter Schmidt, Mercedes Garza, and Enrique Nalda, eds., *Maya* (Mexico: Bompiani, 1998), 243.

40. Linda Schele and David Freidel, op. cit., 418.

41. Ibid., 68.

42. Linda Schele and Peter Mathews, op. cit., 113–114.

43. Douglas Gillette, op. cit., 141.

44. J. E. Cirlot, op. cit., 309–310.

45. Linda Schele and David Freidel, op. cit., 90.

46. Linda Schele and Peter Mathews, op. cit.

47. Douglas Gillette, op. cit., 40.

48. Anthony S. Mercatante, *Who's Who in Egyptian Mythology,* 2nd ed. (New York: Barnes & Noble, 1995), 113–114.

49. Fernando Elorrieta Salazar and Edgar Elorrieta Salazar, *Cusco and the Sacred Valley of the Inca* (Cusco: Peru, 2003), 93. Not only do the Inca equate mummies with trees, the entire city of Ollantaytambo is in the shape of a mythic Tree of Life (see Fernando Elorrieta Salazar and Edgar Elorrieta Salazar, page 95).

50. Douglas Gillette, op. cit., 25.

51. Ibid., 24.

52. "The Death of Mahishasura" http://amsterdam.park.org:8888/India/Durga/dp_myth3.htm.

53. Douglas Gillette, op. cit., 31.

54. Ibid., 92.

55. Ibid.

56. J. E. Cirlot, op. cit., 203.

57. Marius Schneider, born in 1903 in Hagenau, Germany, studied music and philosophy, graduating with a doctorate in philosophy in 1930. During World War II he was a soldier in the German army. He wrote five books, over a hundred articles, and founded a musical institute in Spain in 1944. At the time of his death in 1982, he left behind more than forty years' worth of work on his study of cosmogony (the origins of the universe). He made symbolism a major study in his life, paying

much attention to the symbolism of duality and acoustic symbolism throughout the world. Author J. E. Cirlot quotes Schneider's work extensively throughout his own *A Dictionary of Symbols*. See also Rudolf Haase and Hans Erhard Lauer, introd. by Joscelyn Godwin, *Cosmic Music: Musical Keys to the Interpretation of Reality* (Rochester, Vt.: Inner Traditions, 1990), 33, 34, which includes essays by Marius Schneider.

58. J. E. Cirlot, op. cit., 117.

59. Ibid., 204.

60. Ibid.

61. Ibid.

CHAPTER 11

The 2001 Face on Mars

The M16 Face

On January 31, 2001 Malin/NASA/JPL released seven additional Cydonia images; these had been taken after the images released in April 2000. Among this latest collection of Cydonia images was a fine-detail photograph of the northwestern portion (humanoid side) of the Face, labeled M16–00184 (Figure 11.1). The new "M16 Face" is the highest-resolution picture ever taken of this controversial structure, and we feverishly began to analyze it. Although disappointing in its lack of totality, this image presents details at 1.7 meters, or 5.6 feet, per pixel and is quite revealing.

The new image confirms the existence of the headdress, the triad-leaf crown emblem, the Teardrop feature, and an eye. The image also captures a tiny corner of the mouth. In the forehead area, notice the oval "gemstone" marking a portion of the half emblem that forms the triad-leaf symbol. As mentioned, when mirrored, this half emblem completes a W-shaped, triad-leaf, crown emblem in which the gemstone acts as ornamentation. This Mesoamerican cultural marker has been a consistent feature in all of the last four images of the Face on Mars since 1998.[1] The most positive result of this new image is that it captures the brow area, complete with confirmation of an almond-shaped eye. The central bulge forms a pupil, while

11.1 M16 Face (from the Mars Orbital Camera, M16–00184).

Image and enhancement courtesy of Keith Laney

11.2 M16 Face (positive reversal). Note the projecting edge of the brow and the pronounced eye.

the surrounding depression creates an iris (Figure 11.3). Below the eye is an odd geometric or latticework pattern. This effect is possibly created by the exposure of underlying structural supports. The famous Teardrop feature on the cheek, directly below the eye, appears to have more internal structure in this new image. This is the hand of Isis, as we discussed earlier, on the inverted Face. Here the palm looks steeper and the fingers more rectangular.

Another pleasant surprise on the M16 image is an ornamental "trophy head" mounted just above the eye. Reminiscent of the hidden stags found in the Mexican Huichol Indian painting discussed in Chapter 3 (Figure 3.16), the deer bust forms the brow area of the humanoid side of the Face.[2] An analytical drawing of these fascinating facial features is provided in Figure 11.4.

The symbolic significance of the deer effigy placed above the eye is not clear at this point; however, it may emphasize the idea of sight. Throughout the Middle East, stags and gazelles were used as emblems to denote sharp eyesight.[3] The presence of the stag also reminds us of the many epithets given to the Sumerian water god Enki, who was also known as the

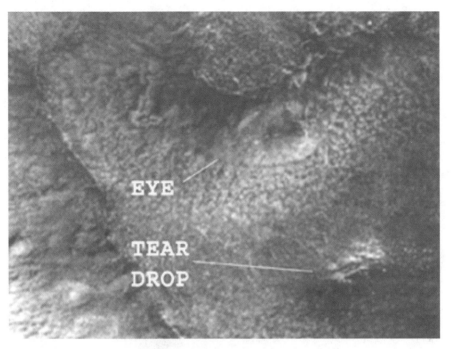

11.3 Detail of the Eye in the New M16 Face. Note the deer ornament in the brow, the almond-shaped eye socket, and the iris in the eye.

Stag of Abzu. A stag can be seen sitting at the feet of Enki on a cylinder seal, illustrated in Chapter 6 (Figure 6.8). In the Maya culture the stag is also seen as a water sign and a symbol of drought.[4]

NASA "Lets the Cat Out of the Bag"

On May 24, 2001, without any fanfare or public notice, NASA released a long-awaited, high-resolution, overhead view of both sides of the Face on Mars[5] (Figure 11.5). Right out of the box NASA presented the new image of the Face as unremarkable and an affirmation that it had finally hit the mark and "scotched this thing for good." According to NASA, the current image clearly maintained their long-held position that this mesa had no resemblance to a human face! The *New York Times* reported: "NASA released

11.4 Analytical drawing of facial features in the new M16 Face. Note the deer ornament above the almond-shaped eye socket.

Drawing by George J. Haas

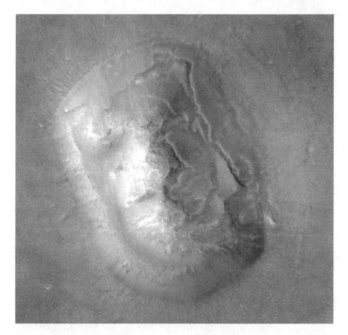

11.5 The April 8, 2001 full-faced image of the Face on Mars (NASA/JPL MOC E03–00824). Note the distinct feline features on the right side.

Courtesy of Keith Laney

a new image [of the Face on Mars] that shows the area in far sharper detail, but reduces any resemblance to a human face."[6]

As soon the new image hit the web, every attempt to mirror the new 2001 Face image was cropped either too wide or too narrow. Many advocates of a bifurcated Face who mirrored the feline side of the geoglyph totally disregarded any sense of a central axis by including portions of the triad-leaf symbol (the W) from the humanoid side.

It has been our experience that all of these bifurcated geoglyphs have precisely placed demarcation lines that are signaled by markers that establish the line of symmetry. The Face has three notable markers that establish a demarcation line. The first runs along the edge of the half emblem of the triad-leaf symbol identified at the center of the forehead on the humanoid side of the Face. A second marker sits at the edge of the central tooth feature on the humanoid side and the protruding tongue and fang on the feline side. The last is a small vertical bar located at the edge of the mane, placed between the feline and humanoid sides of the Face.

You will find this marker, which appears as two vertical parallel lines, in many of the incorrectly mirrored splits of the feline side of the Face. This vertical bar denotes the central axis (or marker) between the two half faces. It should also be noted that because of the three-dimensional aspect of this bifurcated facial structure, the curvature

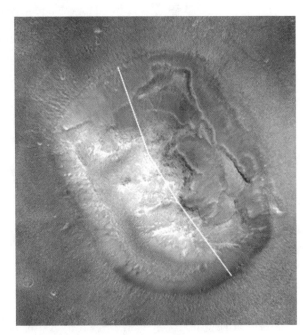

11.6 The line of demarcation (splitting the 2001 Face).

Courtesy of Keith Laney

of its features and the camera angle the demarcation line would appear irregular as it arches across this anthropomorphically designed topography. To help alleviate this distortion, Keith Laney produced an angle-adjusted version of MOC image E1704014 (Figure 11.6).

Narrowly Human

When the new image is mirrored along this central axis, two distinct and separate masks are once again revealed. Although the new image of the Face is highly reflective, the left side still maintains the same humanoid appearance seen in the 1998 image (Figure 11.7B). In the new image we can readily see the now familiar triad-leaf emblem, the eye with deer effigy, a slightly parted mouth, teeth, and nose ornament (Figure 11.7A).

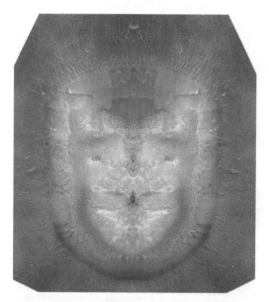 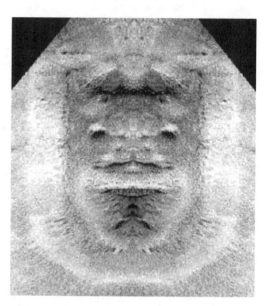

11.7A 2001 humanoid side. Notice the overall highly reflective surface and the distortion of the facial features. Notice also the narrow sides of the fanned, lateral element of the headdress and the triad W emblem.

Courtesy of Keith Laney

11.7B 1998 humanoid side. Note the detail in the facial features—eyes, nose, and chin ornaments.

When the 1998 and 2001 images of the humanoid side of the Face are compared, the luminous, almost translucent condition of the new image becomes quite evident (Figures 11.7A and 11.7B). The cause of this remarkable brilliance may be the effect from the material used to create the facial ornamentation in and around the nose area.

Besides revealing the Face in greater detail, the left side appears slightly compressed. This is due to the foreshortenings caused by the camera angle, which favored the right side. Notice the narrow angle of the base or platform that forms part of the fanned headdress on the left side of the Face. This effect has prompted many researchers to say that the new image has a simian appearance, while the 1998 image was more humanoid. Although similar, the 1998 portrait leans more toward expressing the features of a Neanderthal head. Notice the deep-set eyes, heavy forehead, and the jutting facial structure. There is even evidence of a protruding nasal cavity and prominent jaw line.

The Bearded Jaguar (with Flailing Tongue)

The right side of the 2001 Face image officially confirms the feline aspect of the eastern side of the mesa, as seen in the 1998 image (Figure 11.8B). A comparison to the 1998 image is provided for confirmation that the feline appearance is shared by both exposures (Figure 11.8).

Notice that in the current image of the feline side of the Face, the mask still retains the crown and a square shape to the head. We still see the boxed ears incorporated into the crown, rectangular, squinting eyes, the circular muzzle, and a short, zigzagging ridge line that forms the mane (or beard). Also present is a fang located at the top of what appears to be a flailing "crowned tongue" (Figure 11.9A). When mirrored, this tongue is highly ornamented and has a small face-like visage in the center (Figure 11.9B).

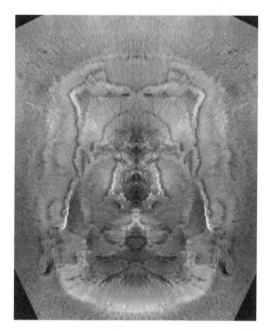

11.8A 2001 feline side (NASA/JPL: E03–00824). Note the crown (terrace) with boxed ears, the square-shaped head, the rectangular squinting eyes, the circular muzzle, and the short, zigzag-shaped mane (or beard). Also note the four little fangs at the top and bottom of the flailing tongue.

11.8B 1998 feline side (negative contrast, SPI–22003). Note that although this side was narrow and dark in the 1998 exposure, causing the foreshortening of its features, its feline aspects are present.

In Mesoamerican cultures the flailing tongue was commonly seen as a sign of bloodletting,[7] and many glyphs can be found depicting jaguars with flailing tongues (Figure 11.10A). The shape of the tongue was synonymous with the sacrificial knife, and in many Maya and Aztec images a protruding tongue took the shape of a knife blade[8] (Figure 11.10B).

As we saw in Chapter 9, the act of bloodletting was one of the most sacred rituals practiced among the cultures of Mesoamerica. The presence of an elaborate crowned tongue on the feline side of the Face is not only significant for its ornamental endowment; it signifies it as a holy repository of sacred blood. To the Maya, the act of bloodletting from the tongue provided the gods with a sacramental vessel of our most precious possession—

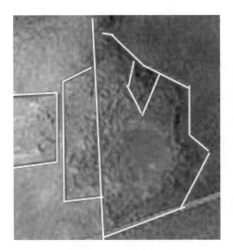

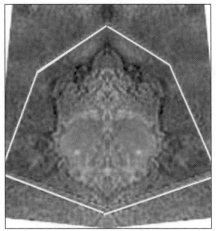

11.9A Crowned tongue (detail outlined in white). Chin and tongue features (divided by the line of demarcation). Note the crowned tongue with fang on the right (the feline side) and the ornamentation on the left (the humanoid side).

11.9B Crowned tongue (detail outlined in white). Mirrored crowned tongue. Notice that the crowned tongue with fang forms a face.

11.10A Flailing tongue. Maya glyph: Jaguar with jeweled tongue.

Drawing by George J. Haas. (Image source: *Maya Hieroglyphic Writing,* by Thompson, Figure 36, #4.)

11.10B Flailing tongue. Aztec sacrificial blade (in the shape of a tongue). Note the small face painted on the blade.

Drawing by George J. Haas. (Image source: *Moctezuma's Mexico: Visions of the Aztec World,* by Carrasco and Matos Moctezuma, page 114.)

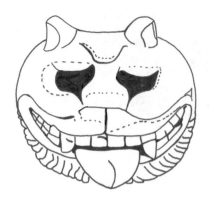

11.11 The Bearded Jaguar (Olmec mask). Note the boxed ears, the protruding tongue, fangs, and the short beard or mane.

Drawing by George J. Haas. (Image source: *Art of Ancient America*, by Disselhoff and Linne, 1966, page 75.)

our DNA. The feline side of the Face also features a small zigzag-shape mane or beard along the neck area. This is another cultural marker consistent with Olmec and Maya iconography, signifying what archaeologists call the Bearded Jaguar. On an Olmec mask from Guerrero,[9] not only do we find the same gesture of the flailing tongue, there is also evidence of a small beard or mane adorning the lower jaw and neck area (Figure 11.11).

The Best of Both Faces

When documenting or archiving a work of art (for a museum or gallery), normally you would have a photographer take three different views of every sculpture to be catalogued. for a bifurcated mask, you would initially take a head-on view, and then a three-quarter view from both the left and right sides. Conveniently, NASA has provided us with such a bracketed set of images of the Face on Mars (Figure 11.12).

The 1976 *Viking* image, although slightly from our left and unevenly illuminated, provides a decent overhead shot. Next, the 1998 image is actually a beautiful image of the humanoid side of the Face, while the feline side (which still bears a remarkable resemblance to a lion) is dark, narrow, and foreshortened. The recent 2001 Face shot provides us with just the opposite view. The new image captures the feline side perfectly, while the humanoid side is narrowed and highly reflective.

What we have done to highlight the best sides of the Face is to crop the last two images at the central demarcation line (Figure 11.6). We have then pasted the best halves together, creating a composite image of the Face (Figure 11.13)—the best of both faces.

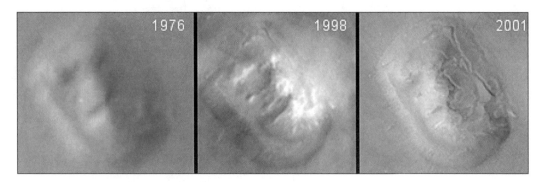

11.12 The three faces.

LEFT, 1976–*Viking* spacecraft (70A13).

CENTER, 1998–*Mars Global Surveyor* (MOC SP1–22003).

RIGHT, 2001–*Mars Global Surveyor* (MOC E03–00824).

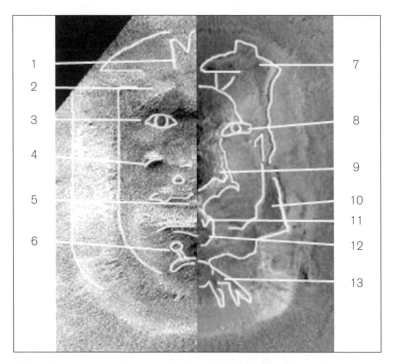

11.13 The best of both faces, 1998 and 2001 (composite image outlined and numbered from 1 to 13). The left side is a crop of the humanoid side of the 1998 (SPI–22003) image. The right side is a crop of the feline side of the 2001 (E03–00824) image. For key to numbers 1–13, see main text.

The image in Figure 11.13 is a highlighted, outlined, and numbered graphic study of a composite image of the Face on Mars. The left half of the composite is a rectified, negative contrast of the 1998 MOC image of the humanoid side. The right side is a negative contrast view of the 2001 image of the feline side.

Below is a concise list of the acknowledged facial features found on both sides of the Face on Mars (as identified by the authors). Just follow the numbers, which correspond to each of the Mesoamerican motifs found in the Face on Mars.

The Humanoid Side:

1. The triad-leaf emblem (one half of the W symbol), located at the center of the headdress
2. Deer effigy, located in the brow area above the eye (see detail of M16 Face)
3. The eye
4. The Teardrop feature (part of the overall nose ornamentation)
5. The tooth (a square-shaped tooth with a circular dental implant at its center
6. Chin ornament (typical of Mesoamerican cultures)

The Feline Side:

7. Crown headdress, referred to as the "terrace" (including the boxed ear)
8. Squinting, rectangular Olmec-shaped eye
9. Nose and muzzle area
10. Zigzag mane or beard
11. Fang
12. Ornamented tongue
13. Avian tag glyph (note the spread wings and legs, while the bird's head is formed by the flailing tongue)

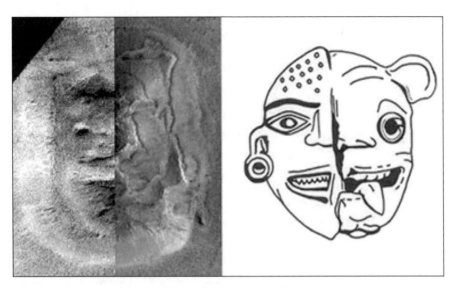

11.14 Two-faced mask comparison (Mars and Earth).

LEFT, Mars (Cydonia) (the best of both faces). Note the humanoid face on the left and feline face on the right (with protruding tongue).

RIGHT, Earth (Mesoamerica) (reversed presentation). Note the human face on the left and the feline face on the right (with protruding tongue).

Drawing by George J. Haas. (Image source: The International Museum of Ceramics in Faenza, Italy.)

When the composite version of the Face is compared to a typical pre-Columbian two-faced mask of a human-jaguar face, the artistic similarities become overwhelmingly evident (Figure 11.14). Notice the protruding tongue on the feline side of the mask. The two half faces are frontal views that are split right down the middle, just like the Face on Mars.

We think it is evident by now that the Face on Mars is not just a "pile of rocks." We agree that the Face on Mars is not a symmetrical human face, and yes, it looks nothing like Elvis—but what it is, is far more important. The Face on Mars is actually a Martian geoglyphic structure that resembles the composite glyphs and two-faced masks of a terrestrial culture that once flourished throughout Mesoamerica.

Notes

1. The triad W-shaped emblem is also present in the most recent *MGS* image of the Face, E10–03730, released in October 2002, and E17-01041, released in April 2003.

2. This odd adornment was simultaneously and independently identified as a "deer effigy" by fellow Mars researcher Terry James (known as KKsamurai).

3. Jean Chevalier and Alain Gheerbrant, *A Dictionary of Symbols* (New York: Penguin Books, 1996), 923. The stag and the gazelle are often regarded as animals with common symbolic attributes, keen eyesight being just one of them. Also see the Bible, Song of Solomon 2:9. In nature, the stag is the victim of the lion, its predator. Notice that a feline face lies on the eastern side of this bifurcated geoglyph.

4. Jean Chevalier and Alain Gheerbrant, op. cit., 282. Another reference to water is found in the inverted pictograph of the Lady of the Jade Skirt on the humanoid side of the Face, discussed in Chapter 5. The duality presented here is clear. The stag signifies drought (no water) while the inverted Lady of the Jade Skirt pictograph signifies the opposite—water.

5. On March 16, 2001 Peter A. Gersten, an attorney-at-law in Sedona, Arizona, sent NASA a "request for data release" letter on behalf of David Jinks, director of the FormalAction Committee for Extra-Terrestrial Studies (FACETS).The letter alleged that NASA not only encouraged but participated in the withholding and misrepresentation of the content of many *MGS* images. On May 24, 2001 NASA released an image of the Face taken almost two months earlier on April 8, 2001.

6. Associated Press, "New View of Mars," *New York Times*, May 25, 2001, A16.

7. Linda Schele and David Freidel, *A Forest of Kings: The Untold Story of the Ancient Maya* (New York: Quill, 1990), 89.

8. Cottie Burland and Werner Forman, *Echoes of the Ancient World: The Aztecs: Fate of the Warrior Nation* (Yugoslavia: Golden Press, 1985), 102.

9. Hans-Dietrich Disselhoff and Sigvald Linne, *The Art of Ancient America* (New York: Greystone Press, 1966), 75. Although this mask is not a typical example of Olmec art, Dr. S. K. Lothrop has ascribed it to the Olmec culture. The mask is from Rio de las Balsas, Guerrero, and exhibits a stylistic influence of the Olmec of La Venta.

CHAPTER 12

The Anunnaki

The Ancient Texts

Many questions arise out of the discovery of these incredible geoglyphic structures on the planet Mars. How did they come to be? Did an intelligent race of beings evolve independently on the planet Mars and then travel to Earth? Was there a former highly advanced race of beings on Earth who traveled to Mars sometime in the distant past and left us these structures? Or is there another explanation that goes far beyond our solar system and even farther into our past? The answer to all of these questions may lie in the oldest known written records of mankind, records that have been found right here on Earth. Over the past 150 years the translation of recovered ancient texts—Sumerian, Akkadian, Babylonian, Assyrian, Egyptian, Hittite, Canaanite—has revealed an incredible story of human history. These texts also provide the story of the times long before man walked the Earth. The volume of these texts is overwhelming. Tens of thousands of clay tablets, found during archaeological digs in the Near East, reveal a whole new chapter of our past.

It is now evident that most of the stories and writings of these cultures had their beginnings in Sumer. Even Biblical tales such as the story of Creation, Adam and Eve, the Garden of Eden, Noah's Ark and the Flood, and the Tower of Babel all found their roots in earlier Sumerian writings. The Sumerian texts are not the oldest, for they themselves make mention of even older writings, "lost books" from before the Deluge. According to these ancient Sumerian texts, there was a time when only the gods were on the Earth—man had not yet been created.

In Chapter 6 we discussed the image found on the City Center Pyramid that we believe represents the Sumerian god Ea/Enki (Figure 6.3). The Sumerian text known as "The Myth of Ea and the Earth" refers to Ea (who became known as Enki) as the leader of a group of fifty travelers who came to Earth from the planet Nibiru (Planet of the Crossing). This happened at a time when man had not yet appeared on Earth. The spelling of the word used in Akkadian for this group of travelers was "Anunnaki," which translates as "the fifty who went from heaven to Earth."[1] In Hebrew and in the Old Testament of the Bible, the Anunnaki were known as the Nefilim, which means "those who were cast down."[2]

At the time of these ancient writings the term "Heaven" did not have the meaning it has today; rather, it was a reference to the area known as the asteroid belt found between Mars and Jupiter. According to the Mesopotamian creation text Enuma Elish, our solar system once housed a feminine planet called Tiamat, which was struck by a rogue planet called Nibiru (later renamed Marduk in Babylonian text). This event happened when the solar system was still forming, and the collision was so severe it caused the splitting of Tiamat. One portion of Tiamat spun off and became the Earth, while the other portion of the planet broke up into small pieces and became the asteroid belt.

Zecharia Sitchin relates:

The Book of Genesis (1:8) explicitly states that it is this "hammered out bracelet" that the Lord had named "heaven" (shamaim). The Akkadian texts also called this celestial zone "the hammered bracelet" (rakkis), and describe how Marduk stretched out Tiamat's lower part until he brought it end to end, fastened into a permanent great circle. The Sumerian sources leave no doubt that the specific "heaven," as distinct from the general concept of heavens and space, was the asteroid belt. Our Earth and the asteroid belt are the "Heaven and Earth" of both Mesopotamian and biblical references, created when Tiamat was dismembered by the celestial Lord.[3]

The Maya have a similar creation story of how the Earth was formed. To create the Earth, the god Quetzalcoatl and his twin, Tezcatlipoca, attacked a celestial monster known as Talaltlecuhtli that was swimming in the primordial waters of the cosmos. They grabbed the monster by the left hand and the right foot and broke it in two. With its upper half they formed the Earth and with its lower half they formed the heavens.[4]

Since the planet Mars is located between the asteroid belt and Earth, the original Sumerian definition of the asteroid belt as Heaven explains Schneider's mountain of Mars—it is the intersection of the circle of Earth and the circle of Heaven. Mars is the mandorla! We quote Schneider once more:

> The mountain of Mars (or Janus) which rises up as a mandorla of the Gemini is the locale of Inversion—the mountain of death and resurrection; the mandorla is a sign of Inversion and interlinking, for it is formed by the intersection of the circle of the Earth with the circle of heaven. This mountain has two peaks, and every symbol or sign alluding to this "situation of Inversion" is marked by duality or by twin heads.

Just as Schneider reveals this idea of duality encoded within the twin-peaked mountain of Mars, with the twin figures of the Gemini and again in the two faces of Janus, we can find common motifs among the iconography of the Maya. The Maya expressed the idea of duality and inversion by producing twin heads in the iconography of their bifurcated masks and composite glyphs.

The Coming to Earth and Mars

The Sumerian story of the coming to Earth and Mars of the Anunnaki is one we have obtained in large part from the writings of noted author and historian Zecharia Sitchin. The story is laid out in his series of books entitled *The Earth Chronicles* and in his latest publication, *The Lost Book of Enki.* This gifted researcher has spent decades translating and interpreting the texts of Earth's most ancient civilizations, as well as chronicling the research and translations of other historians. Although at times his interpretations have received ridicule from some of the scientific community, his claims have continuously been vindicated through new discoveries in science, archaeology, and astronomy.

So the story we are about to relate is not one from our imaginations but from ancient texts, interpreted by those with the knowledge and ability to do so. The following synopsis begins with a short history of the Anunnaki as recorded by the Sumerians and interpreted in large part by Sitchin; it describes the creation of man as well as the existence and possible purpose of the artificial structures left on Mars.

This astonishing story begins on a world known as Nibiru, the Planet of the Crossing, with a conflict between two great leaders who contest the rightful rulership of Nibiru. The ruler was Anu and his challenger was Alalu, who would attempt to usurp the throne. After losing entitlement to the throne, Alalu left in a spaceship in search for gold, which was a highly prized resource on Nibiru. From outside our known solar system, he traveled to the seventh planet, Earth. Alalu contacted his home planet and informed them that he had found the gold for which he had searched. He declared that his discovery would save a dwindling atmosphere on Nibiru, and for this he should be given the throne.[5]

After Anu agreed to another contest with Alalu to determine rulership of Nibiru, Anu's oldest son Ea (or Enki) led a group of fifty Anunnaki to Earth to verify the claims of Alalu. The story and achievements of Ea/Enki are described in one of the longest and best-preserved Sumerian narrative poems so far uncovered, named by scholars the "Myth of Enki and the Land's Order." A portion told in the first person describes his arrival at Earth (some sections unfortunately are illegible):

> When I approached Earth, there was much flooding. When I approached its green meadows, heaps and mounds were piled up at my command. I built my house in a pure place.... My house—its shade stretches over the Snake Marsh ... the carp fish wave their tails in it among the small gizi reeds.[6]

Sitchin has determined that this event took place about 450,000 years ago, during a period when Earth was undergoing an ice age.[7] This date was presented in his book *The 12th Planet,* which was published in 1976 just as NASA was photographing the first image of the Face on Mars. If you recall, it was Richard Hoagland who determined in 1984 that approximately 500,000 years ago (on the Mars Solstice) was the last time that the helical rising of Earth could have been viewed over the Face on Mars. Is it possible that the Face was built to commemorate this major event—the Anunnaki's arrival on Earth?

Searching the planet, Ea determined that sufficient gold could be mined from the area we know as South Africa to meet the needs on Nibiru. To handle the task of setting up mining and processing facilities, more Anunnaki were needed, which led to more divisions of power. To decide who should have control, Anu was summoned to Earth.

Anu was the father of two sons, Ea and Enlil. Although it was Ea, the firstborn, who established the original colony on Earth, Enlil was given rulership over the Earth. Ea was to rule over the seas and oversee the mining operations—a decision that did not sit well with him. The laws of Nibiru were such that an heir had to be not only the oldest son but one conceived by a paternal half-sister.[8] Although Ea was the oldest son, he had not been conceived of a half-sister of Anu. Enlil, however, was, and thus he was considered the legitimate heir. This situation led to a number of confrontations that eventually led to wars between the descendants of Ea and those of Enlil. Ea was not the only one upset with Anu's decisions. Alalu, who was the first to come to Earth and the one who initially discovered gold, was given no title at all. Determined to regain his seat of power on Nibiru, he challenged Anu to the contest he was promised. Anu would once again defeat Alalu, but while celebrating his triumph he was attacked again by Alalu, who bit off and swallowed the testicles of Anu. As punishment for this despicable crime, Alalu was sentenced to exile on the planet Lahmu (Mars). It was here that he would remain alone until his death. Upon arriving at Mars, the pilot of Alalu's spacecraft, Anzu, announced that he would stay with Alalu as his companion and protect him until his death. Anzu said that when the time came he would see to the burial of Alalu, which would be a burial fit for a former Nibiruian king.

It was determined by Anu that, in order to transport the gold from Earth, way stations needed to be set up on both the planet Mars and the small moon of Earth. For his service to Alalu, Anzu would be given command over the way station on Mars. Ninmah, a medical officer who was the daughter of Anu and half-sister to Ea and Enlil, commanded the next flight from Nibiru to Earth. Her job was to transport medical supplies to Earth and begin the process of setting up the way station on Mars. When the spacecraft reached Mars, Alalu was missing and Anzu was found near death.

When he was revived, Anzu told Ninmah of Alalu's death, and led the landing party to the sacred place of his burial. To honor Alalu and to commemorate his deeds as a former Nibiruian ruler and the first king to be buried on an alien planet, the likeness of his face was carved on a large rock. According to Sitchin, Sumerian text recorded the following:

> The image of Alalu upon the great rock mountain with beams they carved. They showed him wearing an Eagle's helmet; his face they left uncovered. Let the image of Alalu forever gaze toward Nibiru that he ruled, Toward the Earth whose gold he discovered! So Ninmah, Exalted Lady, in the name of her father Anu did declare.[9]

Could this ancient text be a reference to the carving of the Face on Mars? Are the remains of the edifice that graced Alalu's tomb the bifurcated mask we see on the Face mesa today? Is Alalu's tomb actually in the shape of a split-faced mask, one half being feline, facing Earth, and the other half being humanoid, facing Nibiru? If true, the tomb of Alalu and the Face on Mars are one and the same, and the settlement that the Anunnaki called the way station is synonymous with Richard Hoagland's City complex, located on a vast plain we call Cydonia.

The Settling of Earth and Mars

The Sumerian texts describe how, over time, many more Anunnaki arrived on Earth,[10] until they numbered 600:

> Assigned to Anu, to heed his instructions, Three hundred in the heavens he stationed as a guard: the ways of Earth to define from the Heaven: And on Earth, Six hundred he made reside, After he all their instructions had ordered, to the Anunnaki of Heaven and of Earth he allotted their assignments.[11]

Sumerian texts chronicle Anu's declaration that the 600 inhabitants of

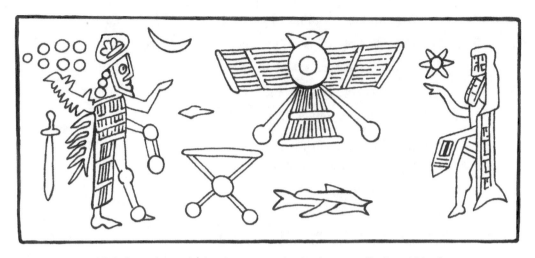

12.1 Sumerian seal (showing communication between Earth and Mars).
Drawing by George J. Haas. (Image source: *Genesis Revisited*, by Sitchin, Figure 91, page 267.)

Earth would be known as the Anunnaki, while the 300 inhabitants of the planet Mars (Lahmu) would be called the Igigi. Figure 12.1 shows a Sumerian cylinder seal bearing two figures. The one dressed in the "Eagleman" or astronaut costume is located near Earth (left), while a "Fishman" (symbol of Ea followers) stands next to Mars (right). An interplanetary vehicle or communications satellite is seen in the middle. The seven circles above the Eagleman, along with the crescent sign, symbolize the Earth and its moon, while the six-pointed star beside the Fishman represents the planet Mars. As these new explorers approached our solar system from outside, Mars is encountered as the sixth planet while Earth is the seventh.

After much time had passed, Anzu and the Igigi on Mars were given passage to Earth for a respite. During this visit a plan was conceived for Anzu and the Igigi to overthrow Enlil's rulership on Earth. Their plan was foiled, however, when Anzu was defeated in battle by Enlil's oldest son and heir, Ninurta. For his evil deed, Anzu was sentenced to death; his body was to be buried on Mars, just as Alalu's had been. Control over the Igigi and the way station on Mars was then given to Marduk, Ea's oldest son and self-proclaimed heir.

The Creation of Man

As time passed, the burden of mining on Earth became too much for the Anunnaki workers and a rebellion threatened. Texts such as the Babylonian "The Creation of Man by the Mother Goddess" and the Sumerian "When the gods as men" spelled out, in detail, the decision to create a primitive worker and the process by which it was to be accomplished.[12] To serve the needs of the Anunnaki, a "lulu" (the mixed one) was created. It is believed that the first man was the result of the genetic manipulation of Anunnaki DNA with that of an existing Earth creature, probably homo erectus.[13] The Creation of Adamu (Earthling) was the work of Ea (Enki) and his half-sister, Ninmah, the chief medical officer (also known as Ninharsag or Ninti). To begin the process, Ea announced: "Blood will I amass, bring bones into being."[14]

The Sumerian text "When the gods as men" then goes on:

While the Birth Goddess is present, Let the Birth Goddess fashion offspring. While the Mother of the Gods is present, Let the Birth Goddess fashion a Lulu: Let the worker carry the toil of the gods. Let her create a Lulu Amelu, Let him bear the yoke....[15]

It was Enki's wife who was chosen to carry the first hybrid child, conceived by Ninki. The text continues:

The newborn's fate thou shalt pronounce: Ninki would fix upon it the image of the gods: And what it will be is "Man."[16]

And Elohim (deities), as written in Genesis 1:26, said: "Let us make Man in our image, after our likeness."

The Maya tell a similar story, wherein a set of heroes on Earth asked their mother to create a race of men who would serve them. They asked the god of the Underworld for a bone of past men to mix with their own blood, which would create a man and a woman to multiply and populate the Earth.[17]

The original humanoid that was created by the Anunnaki, Adamu, was a sterile hybrid and unable to procreate.[18] These hybrids worked as miners in South Africa under the control of Ea and as gardeners in Edin (modern-day Kuwait) under the control of Enlil. The addition of the sex genes by Ea (whose symbol, as you remember, is the helical or entwined serpents) allowed man to procreate. This angered Enlil greatly, and he expelled man from Edin. Left on his own, man began to proliferate at an unexpected rate.

Ea loved to sail the waters and during his travels he came upon two hybrid females. Overcome by their beauty, he mated and impregnated both of them. One bore him a male child, Adapa (the Foundling), and the other a female, Titi (One with Life). The children were raised by Ea and his wife, Ninki. Adapa was taught by Ea himself and became the first civilized man. Adapa and Titi eventually had children themselves, twin brothers Ka-in and Abael;[19] over time, this new breed of civilized man spread throughout the globe.

The Flood

Meanwhile, back on Mars the Igigi were becoming restless and desired the amenities the Anunnaki had on Earth. When it was learned that Marduk was to marry an Earthling, they too wished the companionship of the Earth females. A plot was hatched for 200 of the Igigi to attend Marduk's wedding on Earth and abduct the human females for their wives. They seized the women and, since life on Mars had become too harsh and undesirable, they demanded to be allowed to marry and live on Earth. Marduk and the others agreed.

As a result of this new decree, Enlil became extremely outraged. He had been angered when the first Earthlings were created, furious when they were allowed to reproduce, and now was even angrier at the interbreeding between the Anunnaki and Earthlings that was taking place. This story can be found in the Bible as well, in Genesis 6:1–4:

> And it came to pass, when the Earthlings began to increase in number upon the face of the Earth, and daughters were born unto them, that the sons of the deities saw the daughters of the Earthlings that they were compatible, and they took unto themselves wives of whichever they chose. [Genesis 6:1–2]

> The Nefilim were on the Earth in those days—and also afterward— when the sons of the deities went to the daughters of men and had children by them. They were the heroes of old, men of renown. [Genesis 6:4]

Enlil's anger did not diminish. When it was learned that a catastrophe was approaching the Earth that would cause a worldwide flood, he decided that mankind should not be warned, but perish. Fortunately for mankind, Ea would not allow his creation to be destroyed and he notified one of his sons by an Earthling, Zuisudra, that he must prepare a ship that could withstand the Deluge. In "The Epic of Gilgamesh," the Akkadian text has Ea speaking clandestinely to Zuisudra:

> Man of Shuruppak, son of Ubar-Tutu: Tear down the house, build a ship! Give up possessions, seek thou life! Forswear belongings, keep soul alive! Aboard ship take thou the seed of all living things. That ship thou shalt build—her dimensions shall be to measure.[20]

The legend of the Deluge is found in most cultures and religions of the world, including those of the Americas. The Olmec, Maya, and Inca all believed that they lived in the Fifth Age of the Sun, and that the Fourth Age ended with a universal flood.

After the Deluge (possibly around 11,000 B.C.), when it was discovered that mankind had survived the calamity, Enlil decided that it was destiny, and the will of the "Creator of all things" that man should live. This being the case, the Anunnaki made the decision to endow mankind with the knowledge of farming and animal domestication.[21] Eventually kingship was bestowed upon man. The first human king took the throne in Sumer around 3800 B.C., subsequently in Egypt around 3100 B.C., in the Indus Valley around 2800 B.C., and finally in Peru around 2600 B.C.[22]

Before this time it was only the Anunnaki and their descendants who ruled Earth's city-states. The Sumerian texts state that eight kings reigned during the 241,200 years before the Flood. The Greek historian Herodotus wrote that in the first dynasty of Egypt seven great gods ruled for 12,300 years:

> Ptah (Ea)—9,000 years Ra (Marduk)—1,000 years Shu—700 years Geb—500 years Osiris—450 years Seth—350 years Horus—300 years[23]

The second dynasty of the gods, according to the Egyptian historian Manetho, consisted of twelve divine rulers who ruled for 1,570 years. The first of these was Thoth (Ningishzidda). A third dynasty that lasted 3,650 years was ruled by thirty demigods who were half god, half man—that is, children of the Anunnaki by Earthlings.[24]

Quetzalcoatl and the Mars Beast

Ningishzidda/Thoth was the sixth son of Ea/Ptah and the one to whom Ea taught the sacred knowledge of the ancients, including the science of genetics. Marduk, the first-born son and legal heir of Ea, was so outraged that his brother Ningishzidda was given these secrets and he was not, that he vowed to retaliate. Ningishzidda/Thoth ruled over the lands of Egypt and, like his father, adopted the twin serpents as a symbol. The caduceus was Ningishzidda's emblem, an adaptation of which is still used today as the symbol for medicine (Figure 9.21).

It was around the time that kingship was being presented to mankind in Egypt that Marduk returned from exile (possibly on Mars) and deposed his brother Thoth from lordship throughout Egypt. Historians assign an approximate date of 3100 B.C. to the beginning of dynastic rule in Egypt. The theory states that Ningishzidda/Thoth made his way to the Americas, with his Olmec followers, and became known to the Mesoamericans as Quetzalcoatl, the Feathered Serpent. There are a number of clues to justify this theory.

The Maya record 3113 B.C. as the beginning of their calendar, which is not likely a coincidence. It was Quetzalcoatl who bestowed the knowledge of writing and the arts and sciences on the ancient Maya, and therefore the Maya must have absorbed the concepts of a bifurcated writing system whose existence extends to the two-faced geoglyphs on Mars. The central symbol in the Maya culture was the World Tree. The Sumerian name Ningishzidda means Lord of the Tree of Life.[25] His symbol was the twin serpents, as was that of the Mesoamerican god Quetzalcoatl. The sacred number of the Maya calendar was 52, which was Thoth's magical number. Also, Ningishzidda had been taught the science of pyramid building by his father. Sitchin states:

> It is recorded that when Ninurta (Enlil's foremost son) desired that a ziggurat-temple (stepped pyramid) be built for him by Gudea, it was Ningishzidda (Thoth) who had drawn the building plans....[26] The stepped pyramids of Mesoamerica are of course a wonder to all that

view them, and are exactly the type of structure for which Ningish-zidda (Thoth) was noted.

The Quetzalcoatl connection to Mars is also backed by a number of clues. The ancient cultures of Mesoamerica created an assortment of glyphs that represented most of the known celestial bodies. They had glyphs for the Sun, Venus, Mars, the Moon, and even the Earth. The planet Venus, the evening and morning star, was the most studied and calculated of all the celestial bodies. A variety of geometric glyphs have been catalogued that represent this planet. One of the most interesting of these glyphs is one that appears to be cut in half, possibly representing Venus on the horizon (Figure 12.2A). When the Venus glyph is mirrored along the horizontal baseline, a fully symmetrical eight-pointed star is realized (Figure 12.2B). A similar eight-pointed star was also utilized by the Sumerians to represent the planet Venus (Figure 12.2C). In Mesopotamia the Sumerians counted the planets from the outside in, so as one approaches the Sun from the edge of our solar system, Pluto would be the first, Mars would be the sixth, the

12.2A Aztec Venus glyph (Cacaxla).

Drawing by George J. Haas. (Image source: *Archaeology*, vol. 46, no. 6, Nov/Dec 1993, "Rise and Fall of the City of the Gods," by Carlson, page 66.)

12.2B Aztec Venus glyph (mirrored).

12.2C Sumerian Venus star.

Drawing by George J. Haas. (Image source: *The Cosmic Code*, by Sitchin, 1998, Figure 64, page 174.)

12.3A Sumerian six-pointed star (representing Mars). Star disk.

Drawings by George J. Haas. (Image source: Various Sumerian artifacts.)

12.3B Sumerian six-pointed star (representing Mars). Leaf-shaped star.

Drawings by George J. Haas. (Image source: Various Sumerian artifacts.)

Earth would be the seventh,[27] Venus the eighth, and so on. Because Venus was in the eighth position in the solar system, an eight-pointed star signified its numerical placement. In turn, the red planet we call Mars was associated with a six-pointed star (Figure 12.3).[28]

The Maya depicted the planet Mars in an unusual manner, very different to the ideographic symbols and signs used by other cultures to represent the red planet. According to scholars the Maya sign for Mars was not a star or geometric shape, but a pictograph that archaeologists refer to as the Mars Beast.[29] Two examples are illustrated in the famous Dresden Codex (Figure 12.4). A compressed version of the Mars Beast, also known as the Square-Nosed Beastie, can be seen in the skyband that frames the illustrious Lid of Pacal seen in Chapter 3 (Figure 3.1).[30] The various incarnations of the Mayan creature known as the Zip Monster are actually later representations of earlier prototypes referred to as the Mars Beast and the Square-Nosed Beastie. Many identifying attributes of these creatures are shared, as illustrated in Figure 12.5.

12.4A The Maya Mars Beast. Note the serpentine eye and curled fretted nose. Mars Beast glyph (Dresden Codex).

12.4B The Maya Mars Beast. Note the serpentine eye and curled fretted nose. Mars Beast variant head type (Dresden Codex, detail of a partial figure).

Drawings by George J. Haas

Notice the serpentine eye and the fretted nose. The similarities between the two are striking. Prominent archaeologists such as J. Eric S. Thompson and Michael Coe have identified this glyph not only as a representative of the Zip Monster, but also as a variant signifier for the planet Mars.[31]

Another example of the Maya Zip Monster glyph is found in the center of this introductory glyph (Figure 12.6). The central image features a split-faced glyph of the Zip Monster. The dual attributes of the Zip Monster evolved into one of its greatest incarnations, the Feathered Serpent

12.5 Mars signifier (Maya) (Yaxchilan Zip Monster glyph).

Drawing by George J. Haas. (Image source: *Maya Hieroglyphic Writing*, by Thompson, 1971, Figure 22, #13.)

(Quetzalcoatl). When mirrored, the right half shows a frontal presentation of a serpent, while the left half represents a bird. This bifurcated glyph encoding the embodiment of both the avian and reptilian in simple opposition reflects the true essence of the Mars Beast—duality.

This metamorphic Avian Serpent of the Zip glyph is the personification of three vastly different aspects of Mesoamerican mythology. At one point it represents the primeval Mars Beast, or Zip Monster, which

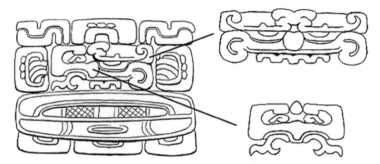

12.6 Maya Zip glyph for the planet Mars. In the center, note a split-faced glyph of a serpent and a bird (the Feathered Serpent).

Drawing by George J. Haas. (Image source: *Introduction to the Study of the Maya Hieroglyphs*, by Morley, 1975, Plate 20, page 221.)

embodies the dual avian and reptilian aspect, echoing back to a time long before Creation. Second, it represents the Feathered Serpent (of Quetzal-coatl fame) designed as a two-faced persona, again representing the two aspects of serpent and bird. Last, it is used as a bearer of celestial numerology for the planet Mars, and, like it did with the Sumerians, it represents "six."

Figure 12.7 shows an Olmec image from a carved blackware bottle found in Mexico that dates to circa 1150 B.C. Notice the duality encoded within this double-headed creature that displays both reptilian and avian features. While the overall shape of the body reflects serpentine aspects, each of the heads with its V-shaped beak displays an avian aspect.[32] These opposing attributes can be viewed as forerunners of the Maya double-headed serpent and the Zip Monster.

One of the heads of the Olmec creature dives to the bottom, or toward the Underworld, while the other rises up above the surface of the primeval waters, reaching up to the heavens. The body of the serpent has eight legs and, most important, it has along its back six diamond-shaped signs that archaeologist Kent Reilly III describes as "stars or celestial symbols."[33] We

12.7 Olmec double-headed serpent with six stars.

Drawing by George J. Haas. (Image source: *Olmec World: Ritual and Rulership*, by Coe, 1996, page 212.)

12.8 The Maya Feathered Serpent with six stars.

Drawing by Linda Schele, © David Schele. (Courtesy of Foundation for the Advancement of Mesoamerican Studies, Inc., www.famsi.org)

wholeheartedly agree with Reilly that these six celestial stars are a cosmological reference. Moreover, we have gone even further and pinpointed that reference as signifying Mars, as the sixth planet.

A second example of a Mesoamerican Avian Serpent with star glyphs is presented here in the form of a Mayan representation of the Feathered Serpent. Just as with the Olmec example of the doubled-headed serpent, this representation of the Feathered Serpent has six dots or spherical stars running along its body (Figure 12.8). The Maya numeral system used a simple dot to designate the number 1, and we are presented here with six dots (along the body of this serpent), which again identify Mars as the sixth planet.

Figure 12.9 is a panel from the Lower Temple of the Jaguar at Chichen Itza, featuring the "old god" Pawahtun discussed in Chapter 6 (Figure 6.31). Two Mars Beasts, or Zip Monsters, rest next to him (there are actually two more Mars Beast heads located at the top of the facade). Notice the six spheres (or dots) in the basket below the head of the Mars Beast on the right (Figure 12.10A). These six spheres signify the same encoded numerical placement of the planet Mars as seen within the body of the Feathered Serpent in Figure 12.8. If you look closely, both of these Mars Beasts have zero glyphs for eyes (Figure 12.10B), again denoting the zero time of Creation. Is this panel trying to tell us that our point of origin (or zero time) was on the sixth planet, the planet Mars?

Another reference to the planet Mars being equated with the number 6 is documented in an inscription that records the

12.9 Pawahtun at the day of Creation. The outer facade of the Lower Temple of the Jaguar (Chichen Itza).

Drawing by Linda Schele, © David Schele. (Courtesy of Foundation for the Advancement of Mesoamerican Studies, Inc., www.famsi.org)

ancient "star war" of 744 A.D. between Tikal and a little-known city found in the jungles of Belize known as Wak Kab'nal, in the kingdom of Naranjo.[34] Wak Kab'nal means "Six Earth Place," and was described as the "city of the

12.10A The Mars Beast with diagnostic integers of 1 and 0. Mars Beast with six spheres in a basket (detail of Pawahtun facade). Note the six round stars in the basket and the use of a zero glyph as an eye.

12.10B Maya zero glyph (found in inscriptions).

Drawing by George J. Haas. (Image source: Introduction to the Study of the Maya Hieroglyphics, Morley, 1975, page 93.)

Square-Nosed Beastie,"[35] who was also Naranjo's founder and patron god.[36] A disagreement in the decipherment of hieroglyphic text has the founding of the kingdom of Naranjo taking place either 22,000 or 896,000 years ago.[37] Because of this immense time frame, one would presume that the Naranjo of Mesoamerica is not the same kingdom founded by the Square-Nosed Beastie. Perhaps it is named after an original settlement established long ago at Six Earth Place—Mars.

Hidden in Plain Sight

What our presentation has shown is that the geoglyphic structures on Mars throughout the Cydonia complex reflect cultural and religious beliefs that include ancient and present-day societies found on Earth. And these geoglyphic structures on Mars are presented in a form that is found in a symbolic language that once flourished in Mesoamerica.

One of the more curious aspects of this is that, although the Maya calendar starts in 3113 B.C., the earliest archaeological discoveries in Mesoamerica are Olmec and only date to circa 1500 B.C. The Maya civilization appears to have peaked in the 500–900 A.D. period and the Aztecs followed them. So what does that tell us about the time frame for the construction of these Martian geoglyphs?

Perhaps the iconography and religion of this period in Mesoamerica evolved out of the remnants of an earlier culture that disappeared from the area due to war or natural disaster. What is more likely is that all of the cultures throughout Mesoamerica and South America are connected to the Anunnaki. The common mythologies shared by the cultures of Mesoamerica and the Sumerians were possibly inherited by these New World cultures through the teachings of Ningishzidda, or, as the Aztec called him, Quetzalcoatl. It was probably also through Ningishzidda that the foundation for a two-faced writing system was set down, to encode the sacred knowledge of the gods.

Is it possible that the structures on Mars are the result of the unendorsed workings of Marduk, who secretly recorded the sacred history of man and the Anunnaki across the surface of Mars in the form of geoglyphs? Outraged by his father's refusal to grant him the sacred knowledge of the ancients, perhaps he stole the knowledge that he felt was due him. Then, in the form of geoglyphic structures, he incorporated this sacred knowledge within the structural design of each building, thereby hiding his blasphemy in plain sight.

Is this the answer? Are the structures on Mars just the remnants of gigantic hieroglyphs, recording the story of a race long forgotten? We may never know until we have a chance to go to Mars and investigate these structures first-hand, or perhaps until new archaeological information connecting Mesoamerica with Mars is uncovered here on Earth. In the meantime, a collaborative research program must be established between NASA and the archaeological communities to further evaluate this lost heritage. The Cydonia Institute, along with many other researchers, is continuing the investigation, and strong evidence has been found that additional half and bifurcated geoglyphic structures can be found beyond the confines of the Cydonia area on Mars.

Growing evidence suggests that the planet Mars was the recipient of an elegant opus that archives the lost heritage of humankind. It appears that some areas of Mars have been literally transformed into a sacred book of pictographic icons. The artistically fashioned surface of these areas simulate holographic pages that unfold within the expansive terrain. A matrix of geoglyphic structures and complexes illustrate a text that is only obscured by its covert design. Like the enigma of the original *Popol Vuh,* the mystery behind this Martian Codex also appears to be hidden from the searcher and the seeker, but for those of us initiated into its matrix ... our eyes are wide open.

Notes

1. Zecharia Sitchin, *The 12th Planet: Book I of The Earth Chronicles* (New York: Avon, 1976), 328.

2. Ibid., vii.

3. Ibid., 230.

4. Mary Miller and Karl Taube, *The Gods and Symbols of Ancient Mexico and the Maya: An Illustrated Dictionary of Mesoamerican Religion* (New York: Thames & Hudson, 1993), 167–168. Also Irene Nicholson, *Mythology of the Americas* (London: Hamlyn, 1970), 154.

5. Zecharia Sitchin states that in order for life on Nibiru to be maintained, fine particles of gold had to be disseminated into the atmosphere. Other

attempts to enhance the atmosphere by stimulating volcanic activity had failed. See Zecharia Sitchin, *The Cosmic Code: Book VI of The Earth Chronicles* (New York: Avon, 1998), 42. Also see Sitchin, *The Lost Book of Enki: Memoirs and Prophecies of an Extraterrestrial God* (Rochester, Vt.: Bear & Company, 2002), 29–38.

6. Zecharia Sitchin, *The 12th Planet,* 291.

7. Ibid., 253.

8. The half-sister had to be a paternal half-sister, that is, both with the same father but different mothers. Sitchin notes that experiments in the animal world indicate that female monkeys prefer to mate with paternal half-brothers. This same propensity is found with wasps. See Zecharia Sitchin, *Genesis Revisited: Is Modern Science Catching Up With Ancient Knowledge?* (New York: Avon, 1990), 183. Is it possible there is a positive genetic outcome from this combination that has been lost over time? The Bible indicates that the father of the Semitic peoples, Abraham, also married a paternal half-sister. See King James Bible, Genesis 20:12.

9. Zecharia Sitchin, *The Lost Book of Enki,* 104.

10. Zecharia Sitchin, *The 12th Planet,* 328.

11. Ibid., 327.

12. Ibid., 336–361.

13. Ibid., 341.

14. Ibid., 350.

15. Ibid.

16. Ibid., 351.

17. Cottie Burland, Irene Nicholson, and Harold Osborne, *Mythology of the Americas* (London: Hamlyn, 1970), 155.

18. Zecharia Sitchin, *Genesis Revisited,* 185–189.

19. Zecharia Sitchin, *The Lost Book of Enki,* 178–182.

20. Zecharia Sitchin, *The 12th Planet,* 381.

21. Zecharia Sitchin, *The Wars of Gods and Men: Book III of The Earth Chronicles* (New York: Avon, 1985), 346.

22. As further evidence for Sitchin's timeline, we look to the ruins at Caral in Peru, which have recently been dated to 2600 B.C. This includes the half-faced geoglyph discussed in Chapter 3, Figure 3.27.

23. Zecharia Sitchin, *The Wars of Gods and Men,* 35.

24. Ibid.

25. Zecharia Sitchin, *The Lost Realms: Book IV of The Earth Chronicles* (New York: Avon, 1990), 269.

26. Ibid. With regard to Ningishzidda being equated with Quetzalcoatl (the Feathered Serpent), it should be noted that just as Ningishzidda is reported as "the builder of pyramids" in Mesopotamia, it is Quetzacoatl who, in the guise of the Zip Monster, is referred to as "a builder of mountains" in the *Popol Vuh.* This relationship between the building of pyramids and mountains is equated by the fact that the Maya built their pyramids to reflect the shape of a mountain.

27. Like the Sumerians, the Maya equated the number 7 as an alternative signifier for the Earth. See Michael D. Coe, *Breaking the Maya Code* (New York: Thames & Hudson, 1992), 133.

28. Zecharia Sitchin, *The 12th Planet,* 263.

29. J. Eric S. Thompson, *Maya Hieroglyphic Writing* (Norman, Okla.: University of Oklahoma Press, 1971), 148. (See also Figure 64, numbers 2, 3, 4.)

30. Two examples of the Mars Beast (Square-Nosed Beastie) can be seen in the skyband that frames the Lid of Pacal. The first is found on the left-side border within the fourth box from the top, just below the X sign. The second Mars Beast is on the right-side border within the second box from the top, just below the X sign. Two additional Square-Nosed Beasties, or Mars Beasts, are found at the end of each branch of the World Tree located in the central panel of the Lid.

31. Michael D. Coe, *The Maya,* 6th ed. (New York: Thames & Hudson, 1999), 159.

32. Michael D. Coe, ed., *The Olmec World: Ritual and Rulership* (Princeton, NJ: Art Museum at Princeton University, 1996), 212.

33. Ibid., 213. (See also 27, 42.)

34. Simon Martin and Nikolai Grube, *Chronicle of the Maya Kings and Queens: Deciphering the Dynasties of the Ancient Maya* (London: Thames & Hudson, 2000), 78.

35. Ibid., 79.

36. Personal conversation with Simon Martin at the 21st Annual Maya Weekend, University of Pennsylvania Museum, Advanced Workshop: "Challenging Text from Tikal," April 6, 2003.

37. Simon Martin and Nikolai Grube, op. cit., 70.

Bibliography

Aldrington, Richard and Delano Ames, *The Larousse Encyclopedia of Mythology* (New York: Barnes & Noble, 1994).

Anderson, Donald M., *Elements of Design* (New York: Holt, Rinehart and Winston, 1961).

Angkor, website: http://www.vwam.com/vets/angkor/introduction1.htm.

Arqueologia Mexicana: Olmecs, no. 19, 1996, National Institute of Anthropology and History and Editorial Raices, Mexico.

Bara, Mike, "New Mars Face Image Analysis and Comment Part III," The Lunar Anomalies, 1999: http://www.lunaranomalies.com.

Barnhart, Robert K., *The Barnhart Dictionary of Etymology* (New York: H. W. Wilson, 1988).

Bassie-Sweet, Karn, "Corn Deities and the Complementary Male/Female Principle," Mesoweb: http://www.mesoweb.com/features/bassie/corn/index.html.

Bauval, Robert and Adrian Gilbert, *The Orion Mystery* (London: Heinemann, 1994).

Bernal, Ignacio, *The Mexican National Museum of Anthropology* (Norwich, Great Britain: Jarrold and Sons, 1972).

Biedermann, Hans, *Dictionary of Symbolism* (New York: Facts on File, 1992).

Black, Jeremy and Anthony Green, *Gods, Demons and Symbols of Ancient Mesopotamia* (Austin: University of Texas Press, 1995).

Bridges, Marilyn, *Markings: Aerial Views of Sacred Landscapes* (New York: Aperture, 1986).

Brockhampton Reference, Dictionary of Classical Mythology (London: Brockhampton Press, 1995).

Campbell, Joseph, *The Mythic Image* (New York: MJF, 1974).

Carlotto, Mark, *The Martian Enigmas: A Closer Look* (Berkeley: North Atlantic Books, 1997).

Carrasco, David and Scott Sessions, *Daily Life of the Aztecs: People of the Sun and Earth* (Westport, Conn.: Greenwood Press, 1998).

Casson, Lionel, and the Editors of Time-Life Books, *Great Ages of Man: Ancient Egypt* (Nederland, NV: Time-Life International, 1966).

Cavendish, Richard, *Mythology: An Illustrated Encyclopedia* (New York: Barnes & Noble, 1993).

Chetwynd, Tom, *Dictionary of Symbols* (London: Aquarian/Thorsons, 1982).

Chevalier, Jean and Alain Gheerbrant, *A Dictionary of Symbols* (New York: Penguin Books, 1996).

Cirlot, J. E., *A Dictionary of Symbols* (New York: Barnes & Noble, 1995).

Coe, Michael D., *Breaking the Maya Code* (New York: Thames & Hudson, 1992).

Coe, Michael D., *The Maya,* 6th ed. (New York: Thames & Hudson, 1999).

Coe, Michael D., *The Olmec and Their Neighbors: Essays in Memory of Matthew W. Stirling* (Washington, D.C.: Dumbarton Oaks, 1981).

Coe, Michael D., *The Olmec World: Ritual and Rulership* (Italy: Princeton/Abrams, 1996).

Cyphers, Ann, *Olmecs, Arqueologia Mexicana,* no. 19, 1996, National Institute of Anthropology and History and Editorial Raices, Mexico.

D'Aulaire, Ingri and Edgar Parin, *The Greek Myths* (Garden City: Doubleday, 1992).

Desroches-Noblecourt, Christiane, *Tutankhamen,* 3rd ed. (Boston: New York Graphic Society, 1978).

Diehl, Gaston, *Picasso* (New York: Crown Trade Papers, 1977).

Diehl, Richard A., *The Olmecs America's First Civilization* (London: Thames & Hudson, 2004).

DiPietro, Vincent and Greg Molenaar, *Unusual Martian Surface Features,* 3rd ed. (Glen Dale, MD: Mars Reasearch, 1982).

Disselhoff, Hans-Dietrich, and Sigvald Linne, *The Art of Ancient America: Civilizations of Central and South America* (New York: Greystone Press, 1966).

Dockstader, Frederick J., *Indian Art in America: The Arts and Crafts of the North American Indian* (Greenwich, CT: New York Graphic Society, 1961).

Early Civilization: The Mind Alive Encyclopedia (London: Marshall Cavendish Books, 1977).

Editors of Reader's Digest, *Mysteries of the Ancient Americans: The New World Before Columbus* (Pleasantville, NY: Reader's Digest Association, 1986).

Editors of Reader's Digest, *The World's Last Mysteries* (Pleasantville, NY: Reader's Digest Association, 1978).

Editors of Time-Life Books, *Lost Civilizations: Incas: Lords of Gold and Glory* (Alexandria, VA: Time-Life, 1992).

Editors of Time-Life Books, *Lost Civilizations: Mesopotamia: The Mighty Kings* (Alexandria: Time-Life Books, 1995).

Edwards, Elwyn Hartley, *The Encyclopedia of the Horse* (New York: Dorling Kindersley, 1994).

Fash, William L., *Scribes, Warriors and Kings: The City of Copan and the Ancient Maya* (London: Thames & Hudson, 1991).

Freeman, Michael and Roger Warner, *Angkor: The Hidden Glories* (Boston: Houghton Mifflin, 1990).

Freidel, David, Linda Schele, and Joy Parker, *Maya Cosmos: Three Thousand Years on the Shaman's Path* (New York: Quill, 1993).

Gates, William, *An Outline Dictionary of Maya Glyphs* (New York: Dove, 1978).

Gilbert, Adrian G. and Maurice M. Cotterell, *The Mayan Prophecies: Unlocking the Secrets of a Lost Civilization* (Rockport, MA: Element, 1995).

Gillette, Douglas, *The Shaman's Secret: The Lost Resurrection Teachings of the Ancient Maya* (New York: Bantam, 1997).

Goodman, Frederick, *Magic Symbols* (London: Brian Trodd Publishing House, 1989).

Graves, Robert, *The Greek Myths* (Wakefield, RI: Moyer Bell, 1994).

Hancock, Graham and Santha Faiia, *Heaven's Mirror: Quest for the Lost Civilization* (New York: Crown, 1998).

Hemming, John and Edward Ranney, *Monuments of the Incas* (Boston: Little, Brown, 1982).

Hoagland, Richard C., "Massive Tetrahedral Ruin," The Enterprise Mission, April 24, 1998: http://www.enterprisemission.com/images/mars/tetra.jpg.

Hoagland, Richard C., "Honey I Shrunk the Face," The Enterprise Mission, April 14, 1998: http://www.enterprisemission.com.

Hoagland, Richard C., *The Breakthroughs of Cydonia,* first edition, limited printing (© Richard C. Hoagland, April 16, 1991).

Hoagland, Richard C., *The Monuments of Mars: A City on the Edge of Forever,* 4th ed. (Berkeley: North Atlantic Books, 1996).

Huxley, Francis, *The Dragon* (London: Thames & Hudson, 1979).

Janson, H. W., *History of Art: A Survey of the Major Visual Arts from the Dawn of History to the Present Day,* 17th printing (New York: Harry N. Abrams, 1973).

Juarez, Magda, Elia Sanchez, and Concepcion Rey, *Olmecs, Arqueologia Mexicana,* no. 19, 1996.

Jung, Carl G., *Psychology and Alchemy* (Princeton: Princeton University Press, 1980).

Leonard, Jonathan Norton and the Editors of Time-Life Books, *Great Ages of Man: A History of the World's Cultures: Ancient America* (New York: Time, Inc., 1967).

Lineman, Carl G., *Dictionary of Symbols* (New York: W. W. Norton, 1994).

Longhead, Maria, *Maya Script* (New York: Abbeville Press, 2000).

Marksman, Robert H. and Peter T. Marksman, *The Flayed God: The Mesoamerican Mythological Tradition: Sacred Text and Images from Pre-Columbian Mexico and Central America* (San Francisco: Harper San Francisco, 1992).

Martin, Simon and Nikolai Grebe, *Chronicle of the Maya Kings and Queens: Deciphering the Dynasties of the Ancient Maya* (London: Thames & Hudson, 2000).

McDaniel, Stanley V., The McDaniel Report Newsletter, "SPAR Scientists Present New Cydonia Analysis: Papers presented at American Geophysical Union Conference." 1998: http://www.mcdanielreport.com.

Muscatine, Anthony S., *Who's Who in Egyptian Mythology,* 2nd ed. (New York: Barnes & Noble, 1995).

Miller, Arthur G., *On the Edge of the Sea: Mural Paintings at Tancah-Tulum, Quintana Roo, Mexico* (Washington, D.C.: Dumbarton Oaks, 1982).

Miller, Mary Ellen, *Maya Art and Architecture* (New York: Thames & Hudson, 1999).

Miller, Mary and Karl Taube, *The Gods and Symbols of Ancient Mexico and the Maya: An Illustrated Dictionary of Mesoamerican Religion* (New York: Thames & Hudson, 1993).

Mookerjee, Ajit, *Kundalini: The Arousal of Inner Energy* (New York: Destiny, 1982).

Morley, Sylvanus G., *An Introduction to the Study of Maya Hieroglyphs* (New York: Dover, 1975).

"NASA Administrator's Third Seminar Series Scheduled," *NASA News,* March 13, 1995: http://www.aero.com/news/nasa_press/n950308d.txt.

"New Mars Photos Cast Doubt on Speculation on a 'Face,'" *New York Times,* April 7, 1998, A24.

Narby, Jeremy, *The Cosmic Serpent: DNA and the Origins of Knowledge* (New York: Tarcher/Putnam, 1999).

Nicholson, Irene, *Mexican and Central American Mythology* (New York: Peter Bedrick, 1985).

Nicholson, Irene, *Mythology of the Americas* (London: Hamlyn, 1970).

Otto, James H., Albert Towle, and Trueman J. Moon, *Modern Biology* (New York: Holt, Rinehart and Winston, 1963).

Planetary nomenclature: http://wwwflag.wr.usgs.gov/USGSFlag/Space/nomen/history.html.

Porter, Frank W., III, *The Maya* (New York: Chelsea House, 1991).

Pozos, Randolfo Rafael, *The Face on Mars: Evidence for a Lost Civilization?* (Chicago: Chicago Review Press, 1986).

Ragghianti, Carlo Ludovico and Licia Ragghianti, *Great Museums of the World: National Museum of Anthropology, Mexico City* (New York, Newsweek, Inc. and Arnoldo Mondadori Editore, 1970).

Reed, Alma M., *The Ancient Past of Mexico* (New York: Crown Publishers, 1966).

Reents-Budet, Dorie, *Painting the Maya Universe: Royal Ceramics of the Classic Period,* 2nd printing (Durham, NC: Duke University Press, 1994).

Reuters, Rio De Janeiro, "Australian Features: Oldest Human Fossil in Americas," reprinted *Calgary Herald,* September 21, 1999.

Roberts, Timothy R., *Myths of the World: Gods of the Maya, Aztec, and Incas* (New York: Metro Books, 1996).

Robinson, James Harvey, *History of Civilization: Earlier Ages* (Boston: Ginn and Company, 1937).

Rubin, William, *Pablo Picasso: A Retrospective, The Museum of Modern Art, New York* (New York: The Arts Publisher, 1980).

Salazar, Fernando Elorrieta and Edgar Elorrieta Salazar, *Cusco and the Sacred Valley of the Incas* (Lima, Peru: Ausonia S.A., 2003).

Schele, Linda and David Freidel, *A Forest of Kings: The Untold Story of the Ancient Maya* (New York: Quill, 1990).

Schele, Linda and Mary Ellen Miller, *The Blood of Kings: Dynasty and Ritual in Maya Art* (New York: George Braziller, 1986).

Schele, Linda and Peter Mathews, *The Code of Kings: The Language of Seven Sacred Maya Temples and Tombs* (New York: Touchstone, 1999).

Schele, Linda, *Hidden Faces of the Maya* (Poway, Calif.: ALTI, 1998).

Schmidt, Peter, Mercedes Garza, and Enrique Nalda, *Maya* (Milano, Italia: Bompiani, 1998).

Shenk, Al, *Calculus and Analytic Geometry* (Santa Monica: Goodyear, 1977).

Sitchin, Zecharia, *Genesis Revisited: Is Modern Science Catching Up with Ancient Knowledge?* (New York: Avon, 1990).

Sitchin, Zecharia, *Stairway to Heaven: Book II of The Earth Chronicles* (New York, Avon, 1980).

Sitchin, Zecharia, *The Cosmic Code: Book VI of The Earth Chronicles* (New York: Avon, 1998).

Sitchin, Zecharia, *The Lost Realms: Book IV of The Earth Chronicles* (New York, Avon, 1990).

Sitchin, Zecharia, *The 12th Planet: Book I of The Earth Chronicles* (New York, Avon, 1976).

Sitchin, Zecharia, *The Wars of Gods and Men: Book III of The Earth Chronicles* (New York, Avon, 1985).

Smith, William, LLD, *A Dictionary of the Bible* (Nashville: Thomas Nelson, 1997).

Soisson, Pierre and Janine Soisson, *Life of the Aztecs in Ancient Mexico* (Editions Minerva–Liber, 1987).

Solis, Ruth Shady, Jonathan Haas, and Winifred Creamer, "Dating Caral, a Preceramic Site in the Supe Valley on the Central Coast of Peru," *Science* 292, no. 5517, April 27, 2001.

Spinden, Herbert J., *A Study of Maya Art: Its Subject Matter and Historical Development* (New York: Dover, 1975).

Stephens, John L., *Incidents of Travel in Yucatan,* Vol. 2 (New York: Dover, 1963).

Stone-Miller, Rebecca, *Art of the Andes: From Chavin to Inca* (New York: Thames & Hudson, 1995).

Stierlin Henri, *Art of the Maya from the Olmec to the Toltec-Maya* (New York: Rizzoli, 1981).

Stuart, George E., "Who Were the Mound Builders?" *National Geographic* 142, no. 6, December 1972.

Tarpy, Cliff, "Unearthing a King from the Dawn of the Maya Place of the Standing Stones" *National Geographic* 205, no. 5, May 2004.

Tate, Carolyn, PhD and Gordon Bendersky, MD, "Olmec Sculptures of the Human Fetus," *PARI Newsletter* 30 (Winter 1999); also: www.mesoweb.com/pari/ news/archive/30/olmec_sculpture.html.

Taube, Karl, *The Legendary Past: Aztec and Maya Myths* (Austin:University of Texas Press, 1997).

Tedlock, Dennis, *Popol Vuh: The Maya Book of the Dawn of Life* (New York: Touchstone, 1986).

Temple, Robert, *The Sirius Mystery: New Scientific Evidence of Alien Contact 5000 Years Ago,* 2nd ed. (Rochester: Destiny, 1998).

The Enterprise Mission website: http://www.enterprisemission.com.

Thompson, J. Eric S., *Maya Hieroglyphic Writing* (Norman: University of Oklahoma Press, 1971).

Thompson, J. Eric S., *The Rise and Fall of Maya Civilization* (Norman: University of Oklahoma Press, 1966).

Tompkins, Peter, *Mysteries of the Mexican Pyramids: Dimensional Analysis on Original Drawings by Hugh Herleston, Jr. and Historic Illustrations from Many Sources* (New York: Perennial Library, 1987).

Tozzer, Alfred M., "Landa's Relacion de las cosas de Yucatan," paper of the Peabody Museum of American Archaeology and Ethnology (Cambridge: Harvard University, 1966).

Vaillant, George C., *Aztecs of Mexico: Origins, Rise and Fall of the Aztec Nation* (Garden City: Doubleday, 1941).

Van Flandern, Tom, *Dark Matter, Missing Planets and New Comets: Paradoxes Resolved, Origins Illuminated* (Berkeley: North Atlantic Books, 1993).

Van Flandern, Tom, Meta Research, April 1998: http://www.metaresearch.org.

Van Flandern, Tom, "Preliminary Analysis of April 5 Cydonia Image from the *Mars Global Surveyor* Spacecraft," Meta Research, 1999: http://www.metaresearch.org.

Watson, James D., *The Double Helix* (New York: Atheneum, 1980).

White, Anne Terry, *Lost Worlds: The Romance of Archaeology,* 8th printing, (New York: Random House, 1941).

Wills, Christopher, *The Wisdom of the Genes: New Pathways in Evolution* (New York: Basic Books, 1989).

Wittington, E. Michael, *The Sport of Life and Death: The Mesoamerican Ballgame* (Singapore: Thames & Hudson, 2002).

Woolley, Leonard, *History Unearthed* (London: Ernest Benn Limited, 1963).

Index

About the Authors

GEORGE J. HAAS is founder and premier investigator of The Cydonia Institute, established in 1991. He is a member of the Pre-Columbian Society of the University of Pennsylvania. Mr. Haas is also an artist, art instructor, writer, and curator. He is a member and former director of the Sculptors' Association of New Jersey. He has also authored monographs for various art exhibitions, and had a one-man show at the OK Harris Gallery of Art in New York City. He became interested in the Face on Mars after reading a book on the subject by Randolfo Rafael Pozos in 1991. He lives in New Jersey with his wife, Dr. Amelia Joy Cole, and he has three daughters—Ebony, Avalon, and Zenith.

WILLIAM R. SAUNDERS graduated from the University of Alberta in Edmonton in 1977 with a Bachelor of Science degree in geomorphology. He began work in the petroleum industry in Calgary, Alberta, in 1978 and currently works as a petroleum geoscience consultant in Calgary. He was reintroduced to the Face on Mars in 1991 by Richard Hoagland's book *The Monuments of Mars*. He began looking at the *Mars Global Surveyor* images on NASA's website with their first release in April of 1998. He met George Haas on a website discussion group shortly thereafter.